AF750976

Diagnostics of Plant Diseases

Edited by Dmitry Kurouski

Published in London, United Kingdom

IntechOpen

Supporting open minds since 2005

Diagnostics of Plant Diseases
http://dx.doi.org/10.5772/intechopen.88565
Edited by Dmitry Kurouski

Contributors
Abhijeet Ghatak, Ritesh Kumar, Shikha Pathak, Nishant Prakash, Upasna Priya, Yong Wang, Monika C. Dayarathne, Amin U. Mridha, Xin-Gen Zhou, Dong-Yan Zhang, Fenfang Lin, Eeshan Kalita, Chayanika Chaliha, Yu Lei, Chandan Singh, Deepak Vyas

© The Editor(s) and the Author(s) 2021
The rights of the editor(s) and the author(s) have been asserted in accordance with the Copyright, Designs and Patents Act 1988. All rights to the book as a whole are reserved by INTECHOPEN LIMITED. The book as a whole (compilation) cannot be reproduced, distributed or used for commercial or non-commercial purposes without INTECHOPEN LIMITED's written permission. Enquiries concerning the use of the book should be directed to INTECHOPEN LIMITED rights and permissions department (permissions@intechopen.com).
Violations are liable to prosecution under the governing Copyright Law.

Individual chapters of this publication are distributed under the terms of the Creative Commons Attribution 3.0 Unported License which permits commercial use, distribution and reproduction of the individual chapters, provided the original author(s) and source publication are appropriately acknowledged. If so indicated, certain images may not be included under the Creative Commons license. In such cases users will need to obtain permission from the license holder to reproduce the material. More details and guidelines concerning content reuse and adaptation can be found at http://www.intechopen.com/copyright-policy.html.

Notice
Statements and opinions expressed in the chapters are these of the individual contributors and not necessarily those of the editors or publisher. No responsibility is accepted for the accuracy of information contained in the published chapters. The publisher assumes no responsibility for any damage or injury to persons or property arising out of the use of any materials, instructions, methods or ideas contained in the book.

First published in London, United Kingdom, 2021 by IntechOpen
IntechOpen is the global imprint of INTECHOPEN LIMITED, registered in England and Wales, registration number: 11086078, 5 Princes Gate Court, London, SW7 2QJ, United Kingdom
Printed in Croatia

British Library Cataloguing-in-Publication Data
A catalogue record for this book is available from the British Library

Additional hard and PDF copies can be obtained from orders@intechopen.com

Diagnostics of Plant Diseases
Edited by Dmitry Kurouski
p. cm.
Print ISBN 978-1-83962-515-2
Online ISBN 978-1-83962-516-9
eBook (PDF) ISBN 978-1-83962-517-6

For EU product safety concerns:
IN TECH d.o.o., Prolaz Marije Krucifikse Kozulić 3, 51000 Rijeka, Croatia, info@intechopen.com or visit our website at intechopen.com.

We are IntechOpen, the world's leading publisher of Open Access books

Built by scientists, for scientists

5,300+
Open access books available

131,000+
International authors and editors

155M+
Downloads

156
Countries delivered to

Our authors are among the
Top 1%
most cited scientists

12.2%
Contributors from top 500 universities

CLARIVATE ANALYTICS
BOOK CITATION INDEX
INDEXED

WEB OF SCIENCE™

Selection of our books indexed in the Book Citation Index
in Web of Science™ Core Collection (BKCI)

Interested in publishing with us?
Contact book.department@intechopen.com

Numbers displayed above are based on latest data collected.
For more information visit www.intechopen.com

Meet the editor

Dmitry Kurouski earned an MS in Biochemistry from Belarusian State University, Belarus, and a Ph.D. (Distinguished Dissertation) in Analytical Chemistry from SUNY Albany, New York. After completing a postdoc in the laboratory of Professor Richard P. Van Duyne at Northwestern University, Illinois, Dr. Kurouski joined Boehringer Ingelheim Pharmaceuticals, where he worked as Senior Research Scientist. In 2017, Dr. Kurouski joined the Biochemistry and Biophysics Department of Texas A&M University as Assistant Professor. His research focuses on the nanoscale characterization of biological and photocatalytic systems using tip-enhanced Raman spectroscopy (TERS) and atomic force microscope infrared spectroscopy (AFM-IR).

Contents

Preface

The continuous growth of the world's population requires a continuous increase in food production. This problem can be solved by expanding agricultural territories. However, this approach is limited and destructive to nature. An alternative approach is to enhance farming efficiency. This novel agricultural concept known as digital farming aims to maximize crop yield while simultaneously minimizing the environmental impact associated with farming.

Digital agriculture requires the development of sensors that can be used directly in the field to monitor plant health. This is important because plant diseases can substantially reduce crop yields. Abiotic stresses, such as drought, salinity or nutritional deficiencies, cause far more serious crop losses. Consequently, timely diagnostics of plant stresses can minimize associated crop yield losses.

Over the last decade, many elegant sensing approaches have been developed. They include but are not limited to satellite- and unmanned aerial vehicle (UAV)-based RGB and thermography imaging as well as hyperspectral, infrared, reflectance and Raman spectroscopy. This book demonstrates the potential of these sensing approaches in diagnosing plant diseases. It also critically discusses the limitations of these innovative technologies in plant-pathogen sensing.

Scientists from various backgrounds ranging from plant biology and pathology to engineering and spectroscopy will find this book useful. The book is also a valuable resource for farmers, plant breeders and pathologists interested in innovative farming technologies that can be used to transform agriculture in the United States as well as overseas.

Dmitry Kurouski
Department of Biochemistry and Biophysics,
Texas A&M University,
College Station, TX, USA

Chapter 1

Application of Spectroscopic Techniques in Early Detection of Fungal Plant Pathogens

Ritesh Kumar, Shikha Pathak, Nishant Prakash, Upasna Priya and Abhijeet Ghatak

Abstract

Among the plant pathogens, around 85% of diseases in plants are caused by fungi. Rapid and accurate detection of fungal phytopathogens up to the species level is crucial for the implementation of proper disease control strategies, which were previously relied on conventional approaches. The conventional identification methods have been replaced by many rapid and accurate methods like high throughput sequencing, real-time polymerase chain reaction (PCR), serological and spectroscopic technique. Among these rapid pathogen detection techniques, spectroscopy is a rapid, cost-effective, non-destructive method and does not require sample preparation. Nowadays, visible, infrared and near-infrared rays are commonly employed for pathogen detection. Fluorescence Spectroscopy, Nuclear Magnetic Resonance (NMR) spectroscopy, Fourier Transform Infrared (FTIR) spectroscopy, Attenuated Total Reflection (ATR)-FTIR spectroscopy, Raman Spectroscopy, Matrix-assisted Laser Desorption Ionization Time-Of-Flight Mass Spectrometry (MALDI-TOF MS). Biocontrol fungus-like *Trichoderma* spp. can be detected with the help of MALDI-TOF MS. Fluorescence spectroscopy used fluorescence emanating from the sample and successfully used in the detection of powdery mildew (*Blumeria graminis*). Hyperspectral imaging is an advanced approach which uses artificial intelligence in plant disease detection. This literature discusses briefly about the features of above-mentioned spectroscopy techniques which may impel the general understanding and propel the research activities.

Keywords: diagnosis, fluorescence spectroscopy, fungal plant pathogens, infrared, near-infrared, spectroscopic techniques

1. Introduction

Fungi, bacteria, viruses, nematodes, and parasitic plants cause plant diseases, which result in a complex relationship between the host plant, the pathogen, and the environment. But most plant diseases (around 85%) are caused by fungi. More than 10,000 species out of 100,000 recognized fungal species may cause diseases in plants. The different strain types and the fungal pathogen's formae speciales make detection and identification more difficult, necessitating the use of specialized techniques. For the implementation of proper disease control strategies, rapid and accurate identification of phytopathogenic fungal pathogens up to the species level is critical. For a long time, experts have used their skills and experience to identify crop diseases with their naked eyes. Finding a specialist and approaching them is not only a time-consuming and repetitive task, but it is also a lengthy and expensive procedure that can take a long time, making the disease very difficult to eradicate and time-consuming in the case of large areas [1]. These traditional methods available for phytopathogenic fungi detection and identification are not always very precise along and even time consuming, which can be shown by **Figure 1**. In order to prevent economic yield losses and safe crop production, advanced plant disease diagnosis can provide rapid, accurate, and effective early-stage identification of plant diseases. For their timely control, early detection and recognition of these phytopathogens are essential. The traditional methods of phytopathogenic fungi detection and identification were mainly based on symptoms, isolation and culture, accompanied by morphological observations along with their biochemical analysis [2]. The study of fungal biology and its relationship with the host plant has made considerable progress in recent years, thanks to the advent of modern holistic and high throughput techniques. The new technologies that are essential to detecting fungal diseases and sensor production are focused on spectroscopy and imaging, mass metabolites and volatile profiling. When "plant disease" and "hyperspectral" are used as key terms to scan for in all databases, according to Web of Science statistics, there are 651 related papers from 1990 to 2019 (**Figure 2**).

Spectroscopy, along with other methods, offers a platform for the creation of non-destructive approaches. The study of the relationship between matter and electromagnetic radiation is known as spectroscopy [3]. Spectroscopy was limited to the absorption, emission, and scattering of visible, ultraviolet, and infrared electromagnetic radiation at the end of the nineteenth century. Throughout the twentieth century, the definition of spectroscopy was extended to include other forms of electromagnetic radiation, such as X-rays, microwaves, and radio waves, as well as energetic particles like electrons and ions [4]. It's ideal for plant disease detection tools to be fast, specific to a particular disease, and sensitive enough to detect symptoms as soon as they appear [5]. With rapid analysis, non-destructive methods meet these requirements, as minimal to no sample preparation is needed. Current research activities in agricultural engineering are working on developing certain technologies to establish a realistic method for large-scale real-time observation of diseases under field as well as semi-field conditions. There are several different kinds of spectroscopy techniques that include specific fungal pathogen detection methods (**Table 1**).

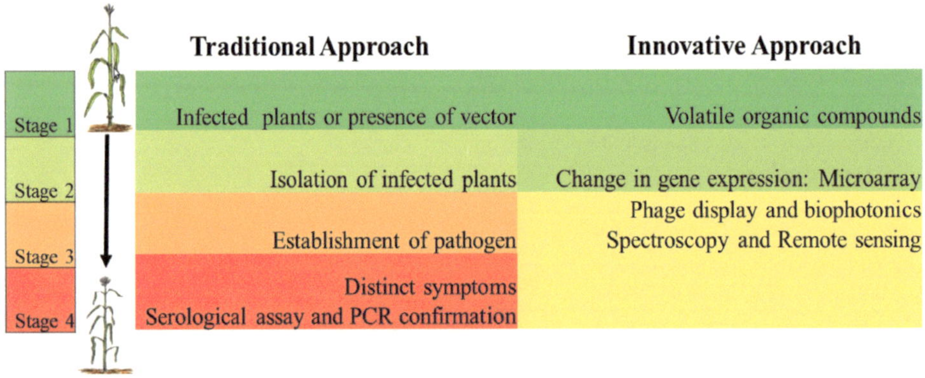

Figure 1.
Comparison between the traditional and innovative approach of plant disease detection by considering four different stages and their timings.

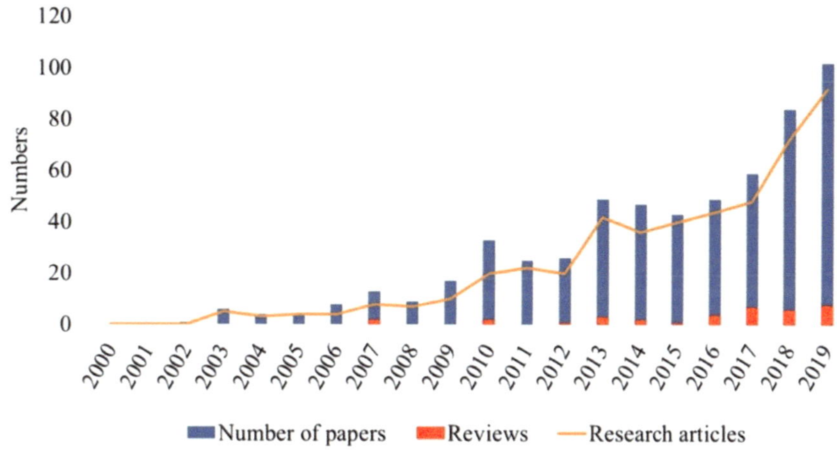

Figure 2.
Number of published articles by year on plant disease with hyperspectral data (adopted from Zhang et al., 2020).

Technique	Crop	Plant Disease	Spectral range	Reference
Visible and Infrared spectroscopy	Wheat	Powdery Mildew and Take all disease	490–540 nm	[6, 7]
	Kiwi	Root rot		[8]
	Orange	Sooty Mold	450–850 nm	[9]
	Grape fruit	Greasy Spot		
	Muskmelon	Powdery mildew		
Fluorescence Spectroscopy	Wheat	Powdery Mildew	blue-to-green (F451/F522); blue-to-red (F451/F687); blue-to-far- red ratio (F451/F736)	[10]
	Barley	Powdery Mildew	410–560 nm	[11]
Nuclear Magnetic Resonance (NMR) Spectroscopy		Mycorrhiza		[12]

Table 1.
Spectroscopy techniques for the detection of fungal pathogens.

2. Plant-pathogen interactions make spectroscopy indispensable

From sowing and growing to harvest, multiple disease-causing pathogens can simultaneously affect plants, reducing the yield and quality of the cultivated plants. It is obvious that many diseases produce similar symptoms and signs on the basis of studies on plant disease detection study, but are caused by very different microorganisms or agents [13]. It can therefore be said that, particularly for non-invasive assay methods, pathogens themselves and plant-pathogen interaction processes are complex. This makes it impossible to use the naked eye or basic machine vision to discriminate against particular pathogens.

2.1 Visible and infrared spectroscopy

As a tool for pathogen detection, non-destructive methods based on visible, infrared and near-infrared spectroscopy are becoming more common as they are fast and cost-effective. In most cases, visible spectroscopy is paired with infrared/near-infrared spectroscopy to detect disease in plants. Dowell et al. [14] used NIR spectroscopy to predict scab, vomitoxin, and ergosterol in single wheat kernels. The application of NIR spectroscopy for mycotoxin measurements in cereals was identified by Pettersson and Aberg [15]. Erukhimovitch et al. [16] explored the ability of FTIR microscopy to differentiate easily and rapidly and to identify different fungi that are responsible for severe agricultural damage. For each of the fungi studied, the findings produced a specific and clear spectral marker. They showed that the spectral region can be regarded as a significant area for simple and accurate differentiation between the different fungi examined, ranging from 1,000 to 1,800 cm^{-1}. Huang and Apan [17] used a portable spectrometer to collect hyperspectral data in the field to detect *Sclerotinia* rot disease in celery and found that adequate reflectance in the visible and infrared ranges from 400 to 1,300 nm produced similar results as the entire spectrum (400–2,500 nm).

2.2 Fluorescence spectroscopy

Fluorescence spectroscopy is a method of electromagnetic spectroscopy that analyses fluorescence from the sample of interest. The sample is excited by using a light beam that results in a lower energy light emission, resulting in an emission spectrum that is used to interpret results [18]. Green leaves generate two forms of fluorescence: blue-green fluorescence (about 400–600 nm range) and chlorophyll fluorescence (about 650–800 nm range). Fluorescence spectroscopy, with a high sensitivity and specificity rate, seems to be a promising diagnostic technique that makes it an ideal diagnostic method. Fluorescence spectroscopy can be used to track food shortages, environmental stress levels, and plant diseases [19]. In four genotypes of spring barley in healthy leaves, as well as leaves, inoculated with powdery mildew pathogen (*Blumeria graminis*), Leufen et al. [11] studied the ability of three optical devices, namely fluorescence lifespan, image-resolved multispectral fluorescence and selected indices of a portable multiparametric fluorescence system for the proximal sensing of plant-pathogen interactions (*Puccinia hordei*). Important variations were found between healthy and diseased leaves.

2.3 Nuclear Magnetic Resonance (NMR) Spectroscopy

The introduction of a higher magnetic field has brought greater sensitivity and spectral resolution. Technology developments have made it possible to combine various NMR techniques that allow for metabolic, anatomical, and physiological knowledge. Another advantage of NMR measurements in researching the biochemistry of mycorrhizas is the ability to spectroscopically distinguish host from fungal metabolites without the need for separation or chemical derivatization. Through the implementation of high-resolution solid-state magic angle spinning nuclear magnetic resonance, intact tissue analysis was possible (HR-MAS NMR). Pfeffer et al. [12] studied the application of Nuclear Magnetic Resonance (NMR) to the two major types of mycorrhiza (ectomycorrhiza and arbuscular mycorrhiza) in order to address the physiological question of the sufficient discrepancy between these two mutualistic symbioses. They found that NMR isotopic labelling can be used to investigate the transfer of substrates between *in-vivo* and *in-vitro* symbionts, as well as the formation of secondary metabolites in response to colonisation. It can also be used to evaluate the locations of biosynthesis and storage compound translocations in mycorrhizal fungi.

2.4 Fourier transform infrared (FTIR) spectroscopy

The Fourier transform infrared spectroscopy (FTIR) is one of the methods that has been successfully used to detect and recognise fungal plant pathogens [20]. This technique was shown in some studies to be capable of discrimination not only at the genus level but also at the species level [21]. The vast majority of these experiments involved both bacteria and fungi. Because of its sensitivity, rapidity, low cost, and simplicity, FTIR spectroscopy has the potential to be a very useful method for detecting and recognising fungal pathogens in agriculture [22].

Erukhimovitch et al. [23] used standard FTIR spectroscopy methods to identify control uninfected potato tubers and tubers naturally contaminated with fungal pathogens. To confirm the absence of fungal infection, samples from uninfected control potatoes were grown in the required growth medium. Thin potato samples were prepared directly from the surface of uninfected and infected potatoes for FTIR analysis. The FTIR spectra of both uninfected and contaminated samples collected from potatoes. Tubers indicate a disparity in spectra between infected and uninfected tissues, with unique clear spectral bands appearing in the spectra of infected tissues.

2.5 ATR-FTIR spectroscopy

Attenuated total reflection Fourier transform infrared abbreviated as ATR-FTIR spectroscopy imaging is a non-destructive imaging method which can be exploited for wide range of samples and system studies. It is highly versatile in nature and can be applied in biomedical sciences and horticulture industries for indentification and interaction of pathogen with host.

ATR-FTIR has been used in characterization of fungal isolates of *Rhizoctonia*, *Verticillium*, *Colletotrichum*, *Fusarium* species [24, 25] and *Geotrichum candida* [26]. Diagnostic analysis and exploratory features of ATR-FTIR spectra offers potential detection of intact host–pathogen systems and other biological insights.

2.6 Raman spectroscopy

Raman spectroscopy (RS) is based on the Raman effect, which states that when incident light (750–850 nm) excites molecules in a tissue, the molecules will reflect light at a different wavelength. The wavelength of the reflectant light is unique to various chemical components, allowing for chemical synthesis to be identified by atheromatous plaque. It may distinguish between various plaque components including elastin, collagen, cholesterol, cholesterol esters, lipids, carotenoids, and calcium apatite deposits. To distinguish normal tissue from abnormal tissue, fluorescence spectra are obtained from a coronary artery by supplying excitation light and collecting emitted light through flexible optical fibers. RS is a nondestructive, label-free spectroscopic technique that offers knowledge about the chemical composition of examined specimens. Food chemistry [27], electrochemistry, forensics and materials science, and agricultural sciences are among its practical applications. Farber et al. [28] demonstrated that using a hand-held Raman spectrometer in conjunction with chemometric analyses, it is possible to differentiate between healthy and diseased maize (*Zea mays*) kernels, as well as between different diseases, with 100% precision. The study was compact and sample-agnostic, implying that it could be retooled and performed autonomously for other crops.

Van Duyne discovered in 1977 [29] that coherent oscillations of an electron cloud at the surface of nanoparticles would amplify Raman scattering by a factor of 108. This phenomenon, known as surface-enhanced Raman spectroscopy (SERS), allows for single-molecule detection and has thus been widely used to detect fungi-related toxins [30].

2.7 MALDI-TOF MS

With technological advancement new and relaible tools are emerging for detection and identification of plant pathogens, one of such technique is Matrix Assisted Laser Desorption Ionization-Time-Of-Flight Mass Spectrometry (MALDI-TOF MS). Identification of several phytopathogenic fungal genera such as *Alternaria*, *Fusarium*, *Monilinia*, *Puccinia*, mildews and potential bicontrol fungus like *Trichoderma* and *Metarhizium* has been successfully done with MALDI system. Although identification of several genera of fungus is done by MALDI-TOF MS, still this technology is not yet considered a standard tool for the fungal identification and their functions (**Table 2**).

3. Some other uses of spectroscopy in plant pathogen detection

Sankaran et al. [42] used spectroscopy imaging technologies and made a distinction between normal healthy and diseased leaves of many plants. These systems have the advantage of being effective in detecting plant diseases. The challenges that these techniques face, include determining the best approach for a specific plant disease and automating techniques for continuous plant disease monitoring.

Ewis Omran [43] demonstrated a method for early detection of plant disease by focusing on the impact of fungal diseases such as leaf spots on peanut. *In-situ*

Fungus	Contribution	Reference
Alternaria	Detection of alternariol, alternariol monomethyl ether, and tentoxin from *Alternaria*	[31]
	Separation of *A. porri, A. dauci, A. tomatophila and A. solani*, from a complex	[32]
	Identification of 60 isolates, among them 12 were *Alternaria* species.	[33]
Fusarium	Identification and characterisation of *Fusarium verticillioides* and fumonisins.	[34]
	Differentiation of various *Fusarium* spp. based on spores	[35]
Downy and powdery mildew fungus	Identification of the *Bremia lactucae* and *Oidiumneo lycopersici*, from infected leave	[36]
Bremia, Oidium	Identification of ribosomal proteins and histones as markers for the biotyping of plant pathogens	[37]
Gibberella	Characterisation of *Gibberella zeae* conidia by on-target trypsin digestion	[38]
Monilinia	Identification of *Monilinia* brown rot fungi from infected fruits	[39]
Trichoderma	Direct identification of hydrophobins in *Trichoderma* isolates.	[40]

Adopted from Drissner and Freimoser [41].

Table 2.
MALDI-TOF MS studies of agriculturally important fungi.

spectroscopy was used to identify early and late leaf indices. Thermal and spectral measurements were also used to differentiate between healthy and contaminated plant leaves. Later, a drop in plant chlorophyll was also observed.

Martinelli et al. [44–46] identified modern nucleic acid and protein analysis-based methods for identifying disease in plants. The authors also identified various mobility spectrometer and lateral flow devices that detect early infections directly on fluid, which summarised remote sensing technologies combined with spectroscopy-based methods and resulted in high spatialization whirlpools.

Ray et al. [47] used hyperspectral reflectance data from a spectro radiometer with spectral ranges of 770–860 nm and 920–1050 nm to detect late blight disease in potatoes and found a significant difference between healthy and diseased plants. Zhang et al. [48, 49] used a spectro radiometer with 32 spectral features and different models to compare normal and contaminated leaves' hyperspectral reflectance. Every model was thoroughly examined using t-tests, correlation analyses, and fisher linear discriminant analysis, and it was discovered that PLSR outperformed the MLR model. FLDA also included accurate information.

At an early stage, Romer et al. [50] proposed a method for distinguishing leaf rust wheat leaf from safe leaf. The authors provided pre-symptomatic identification, which was followed by classification using the support vector machine approach. It also shows how to collect different parameters using fluorescence detection using a fluorescence spectrometer. Further support vector machine was being used for classification for healthy and inoculated leaves.

4. Spectroscopy based methods for remote sensing of plant disease

In remote sensing of plant diseases spectroscopy is among the most used methods which involves imaging or no-imaging sensors, visible wavelength, near infrared wavelength, and shortwave infrared wavelength. These techniques are considered reliable in crop disease monitoring as they are promising in operational instruments, efficacy, cost-efficiency and flexibility.

Under non-imaging spectroscopy approach of remote sensing of plant disease data is recorded based on inherent optical properties of leaf and leaf pigments, structural characteristics and chemical components [51]. Leaf spectra were collected either in field or laboratory to determine spectral regions (visible wavelength, near infrared wavelength, and shortwave infrared wavelength) to detect diseases. Some of the most studied fungal diseases using this method arewheat powdery mildew caused by *Erysiphe graminis* sp. *tritici* and take-all disease caused by *Gaeumannomyces graminis* sp. *tritici* [6, 7]. Apart from fungal diseases this method is also been used in detection of different viral diseases and numerous insect pest incidence in crop fields [44–46].

Nowadays, hyperspectral imaging instruments are being incorporated into monitoring and assessment of plant diseases. Some of the laboratory-based studies for imaging spectroscopy includes head blight and disease Fusarium fungal infection in wheat [52], early stage detection of diseases of sugar beet [53], and detection of sugar beet rust, Cercospora leaf spot and powdery mildew on sugar beet leaves [54]. A large range of statistical methods were applied under these studies for image analysis which includes principal component analysis (PCA), linear regression, support vector machine (SVM) classification and spectral angle mapper (SAM) classification producing high accuracy for detection of disease. Data obtained from both field and airborne hyperspectral were used to assess the severity of Rhizoctonia crown and root rot disease in sugar beet [55] and yellow rust in wheat [56].

5. Hyperspectral imaging

Image analysis is a new breakthrough in the field of plant disease identification and detection. Image analysis has huge potential in near as it is non-invasive and autonomus approach of detecting stress (biotic and abiotic) in plant [57]. It involves extraction of image from the images captured digitally. The image can be captured from a varied source viz., smart phones, digital camera, highly specialized cameras which are designed to extract variety of information from the image. Hyperspectral thermal, Multispectral, 3D sensor, Red Green Blue (RGB) method, Chlorophyll fluorescence are methods used in plant disease detection. Among these, Hyperspectral imaging and RGB is mostly preferred for plant disease identification [58]. In Hyperspectral imaging, camera is capable to capture light wavelength beyond visible range (400–700 nm). Human eye can perceive electromagnetic spectrum ranging from 400 to 700 nm but Hyperspectral imaging ranges from 250 nm (Ultraviolet, UV) to 2500 nm (Short-wave infrared range, SWIR). Camera is combined with some specific sensors to widen the coverage of the capturing spectrum. Usually, certain sub-range of electromagnetic range of radiation is captured by the camera viz., UV (250–400 nm) or visible and near infrared range (NIR) (400–1300 nm) or SWIR (1300–2500 nm). 400–700 nm wavelength are capable to detect changes in pigmentation of leaf while 700–1300 nm are to detect mesophyll cell structure, however extended range of wavelength 1300–2500 nm are needed to analyze content of water in plant.

6. Hyperspectral imaging technology

The image is captured in various way with the help of different hardware approach. The various hardware approaches include push broom, liquid crystal tunable filters, filter wheel others [59]. Among these, push broom technology involves incidence of light on a prism or convex grating leading to formation of narrow wavelength spectrum which further recorded on light sensitive chip (analogous to digital camera). A push broom device comprises of the camera, a lens and a spectrometer. This device involves simultaneous capture of single spatial line and whole range of colour spectrum. After scanning first line, camera moved to capture the next line and final image is formed. Camera act as a line scanner and after completion of scanning final image is formed. Snapshot is an alternative to push broom approach. Instead of providing point-and-click measurements, in Hyperspectral devices onus lies on the developer to develop capture process. Capture of image results in generation of large dataset sets which is further analyzed to obtain useful information. A simple and convenient way is to analyze this large dataset is to consider positions of small number in the captured wavelength. This approach facilitates countering the effects of relative changes in light by taking into account the ratios of data values. This is achieved by combining two or more wavelengths of light which is referred as "indices". In order to interpret the captured data, numerous such indices have been formulated through pre-considered biological reasoning (eg. Knowledge that particular wavelength refers to the specific properties in cell structure) or because of limitations of particular wavelengths obtained from capture equipment (e.g. indices which are developed from data obtained from multispectral remote sensing, may have limited number of wavelength). When these indices are applied to plant material then referred to as vegetation indices. Several such indices (**Table 3**) exist and each indices uses distinct set of wavelength measurements to describe different physiological attributes of plant.

S.N.	Index	Formula	Information	Reference
1	Normalised difference vegetation index (NDVI)	(RNIR − RRED)/(RNIR + RRED) RRED ~680, RNIR ~800	Range: − 1 to 1 Common range: 0.2–0.8 Broadband	[60]
2	Red edge NDVI	(R750 − R705)/(R750 + R705)	Range: − 1 to 1 Typical healthy range: 0.2 to 0.9 Narrowband (hyperspectral data)	[60]
3	Simple ratio index (SRI)	RNIR/RRED RRED ~680, RNIR ~800	Range: 0 to >30 Typical healthy range: ~ 2–8 Broadband	[61]
4	Photochemical reflectance index (PRI)	(R531 − R570)/(R531 + R570)	Range: − 1 to 1 Typical healthy range: − 0.2 to 0.2 Vegetation health prior to senescence	[60, 62]
5	Plant senescence reflectance index (PSRI)	(Red–Green)/NIR	Range: − 1 to 1 Typical healthy range: − 0.1 to 0.2 >PSRI ~ canopy stress, onset of senescence, fruit ripening	[60]
6	Normalise dphaeophytinizationindex (NPQI)	(R415 − R435)/(R415 + R435)	Chlorophyll degradation 0.56–1.41 Unacidified and acidified solutions	[63]
7	Structure Independent Pigment Index (SIPI)	(R800 − R445)/(R800 + R680)	Range: 0–2 Typical healthy range: 0.8–1.8 Good with canopy variety	[60, 62, 64]
8	Leaf rust disease severity index (LRDSI)	6.9 × (R605/R455) − 1.2	Accuracy of 89% in study may vary with other data.	[65]

Table 3.
Vegetation indices, their formulae and information.

Normalised difference vegetation index (NDVI) is widespread and highly popular metrics used to measure the general crop health status [66, 67]. NDVI is used to detect biotic stress due to Sunn pest/cereal pest, *Eurygasterintegriceps Put.* (*Hemiptera: Scutelleridae*) in wheat. There are many specific or disease centric indices which are helpful in detection and quantification of specific disease [61]. Leaf rust disease severity index (LRDSI) is an example of disease centric index having 87–91% accuracy in detection of wheat leaf rust (*Puccinia triticina*) [65].

Red edge approach is another commonly used method where abrupt rise in reflectance at the red/near infrared border is detected. The red edge position comprise of narrow section of electromagnetic spectrum (690–740 nm) where visible light spectrum ends and the NIR starts. This section of wavelength range showed good spectral response for green plant material. Chlorophyll has high absorption capacity and low reflectance for 700 nm light wavelength but it has strong reflection for infrared i.e., light wavelength starting from 720 nm. A red edge based disease index is largely used for detection of powdery mildew of wheat (*Blumeria graminis* f.sp. *tritici*). However, red edge approach has less accuracy than Partial least squares regression (PLSR) method. PLSR has a statistical approach.

7. Classification approach using subset of selected spectrum data

This approach involves subsampling of particular wavelength from the full spectrum. Unlike multispectral data, specific wavelength can be chosen autonomously or manually from any position in the captured wavelength range. Wheat field experiment study involves NDVI response to remove all datasets except from the leaves followed by ANCOVA (Analysis of Co-variance) to detect specific wavelength of band which again followed by quadratic discriminant analysis (QDA) which distinguish the spectra of healthy plant from diseased leaves (yellow rust) [68]. This is the typical operation flow in hyperspectral image analysis. Use of QDA enhances accuracy up to 92% along with four bands [68]. Likewise, several techniques are used in plant disease detection in Hyperspectral imaging technology (**Table 4**).

Multi-layer Perceptron (MLP) approach detect yellow rust in wheat field uses a spectrograph of range 460–900 nm and 20 nm of spectral resolution [71]. The image is captured by the spectrograph by handheld system. Four significant light wavelengths were selected. The 'variable selection' method was employed for the selection of first two wavelengths using discriminant analysis and F-test. Another pair of wavelengths were selected by using NDVI wavelength. Moshou employed a neural network comprising four inputs, two outputs and one hidden layer consist of ten neurons. It has classification accuracy of 98.9% for healthy plants while 99.4%

S.N.	Technique	Plant disease	Accuracy	References
1	89 Quadratic discriminant analysis (QDA)	Wheat (yellow rust)	92%	[68]
		Avacado (laurel wilt)	94%	[69]
2	Decision tree (DT)	Avacado (laurel wilt	95%	[69]
		Sugarbeet (cerospora leaf spot)	95%	[70]
		Sugarbeet (powdery mildew)	86%	
		Sugarbeet (leaf rust)	92%	
3	Multilayer perceptron (MLP)	Wheat (yellow rust)	98.9/99.4%	[71]
4	Partial least square regression (PLSR) Raw Savitsky-Golay 1st derivative Savitsky-Golay 2nd derviative	Celery (sclerotinia rot)	88.92% 88.18% 86.38%	[17]
5	Partial least square regression (PLSR)	Wheat (yellow rust)	92%	[72]
6	Fishers linear determinant analysis	Wheat (aphid) Wheat (powdery mildew) Wheat (powdery mildew)	60% 90%	[48, 49]
7	Erosion and dilation	Cucumber (downeymildew)	90%	[73]
8	Spectral angle mapper (SAM)	Sugarbeet (cerospora leaf spot)	89.01–98.90%	[53]
		Sugarbeet (powdery mildew) Sugarbeet (leaf rust) Wheat (head blight)	90.18– 97.23% 61.7% 87%	[52]
9	Artificial neural network (ANN)	Sugarbeet (cerospora leaf spot) Sugarbeet (powdery mildew) Sugarbeet (leaf rust)	96% 91% 95%	[70]
10	Support vector machine (SVM)	Sugarbeet (cerospora leaf spot)	97%	[70]
		Sugarbeet (powdery mildew) Sugarbeet (leaf rust)	93% 93%	[74]

Table 4.
Techniques used in Hyperspectral imaging for detection plant diseases.

for diseased plants. In machine learning, a highly sophisticated approach known as deep learning is gaining popularity. Deep learning consists of artificial neural network having a structure containing numerous layers. Each layer of neuron implicitly represent features obtained from the data which in turn complex information can be furnished from later layers and whole image features can be obtained from network. Convolutional neural networks (CNN) and Artificial neural networks (ANN) popularly used in deep learning. Using CNN, deep learning is reported to identify 26 diseases in 14 crop species [75]. AlexNet and GoogLeNet are two popular versions of CNNs having accuracy in disease detection up to 97.82% and 98.36% respectively. Both versions use datasets of 54306 images where 80% involve in training and 20% testing.

8. Disease identification

Apart from detection of presence and absence of disease, research is focusing on distinguishing between different disease and identification of specific disease. Spectral information divergence classification is one of the approaches fulfilling this purpose. Comparative analysis is performed between observed spectra and available reference spectra (a library of diverse spectra). Spectral information divergence employed in detection of canker legions on citrus. Greasy spot, melanose, insect damage, wind scar and scab were detected in grape with 95.2% classification accuracy [76].

9. Quantification of disease severity

SAM (Spectral angle mapper) is an approach is used to quantify severity of plant disease. SAM approach matches pixel spectra to available reference spectra leading to classification of pixels. This classification involves calculation of angle between the spectra. These spectra further considered as n-dimensional vectors in the space [77]. This approach has moderate level of success and widely used by researchers. Yuhas et al. [77] recorded the Fusarium head blight severity in wheat using hyperspectral data of range 400–1000 nm and spectral resolution 2.5 nm. In quantification of disease severity, SAM accounts for 88% classification accuracy. Malhein et al. [78] also quantified disease severity of Cercospora leaf spot, powdery mildew, and rust in sugar beet using SAM approach.

10. Conclusion

Early detection of plant disease plays key role in planning of plant disease management programme. Nowadays many non-destructive methods of plant disease detection are gaining popularity. Different spectroscopic methods offer a non-destructive method of plant disease detection. These methods use visible, ultraviolet, infrared and near-infrared lights to capture image of plant sample. Fluorescence microscopy, NMR, FTIR spectroscopy, ATR-FTIR spectroscopy, Raman spectroscopy, MALDI-TOF microscopy, and Hyperspectral imaging are different nondestructive and spectroscopy-based method of plant disease detection. Among all these methods visible, ultraviolet, infrared and near-infrared wavelength of lights are used for image analysis of diseased plant sample. In hyper spectroscopy imaging, image captured using visible, ultraviolet, infrared and near-infrared wavelength of lights were further analyzed by using artificial intelligence. As these methods are very much promising but still their accuracy needs to be improved.

Quantification of disease is also great concern. These methods are not very much promising in quantification of disease severity in plants. Future research must be focused on developing a system which give promising result regarding quantification of plant disease severity.

Author details

Ritesh Kumar[1], Shikha Pathak[2], Nishant Prakash[3], Upasna Priya[4] and Abhijeet Ghatak[4*]

1 Department of Plant Pathology, MSSSoA, Centurion University of Technology and Management, Paralakhemundi, Gajapati, Odisha, India

2 Department of Plant Pathology, Bidhan Chandra Krishi Viswavidyalaya, Nadia, West Bengal, India

3 Plant Pathology, Krishi Vigyan Kendra, Arwal, Bihar, India

4 Department of Plant Pathology, Bihar Agricultural University, Bhagalpur, Bihar, India

*Address all correspondence to: ghatak11@gmail.com

IntechOpen

© 2021 The Author(s). Licensee IntechOpen. This chapter is distributed under the terms of the Creative Commons Attribution License (http://creativecommons.org/licenses/by/3.0), which permits unrestricted use, distribution, and reproduction in any medium, provided the original work is properly cited.

References

[1] Mishra, P., Polder, G. and Vilfan, N. (2020). Close range spectral imaging for disease detection in plants using autonomous platforms: a review on recent studies. Curr. Robot. Rep. 1, 43–48.

[2] Tan, D.H.S, Sigler, L., Gibas, C.F.C. and Fong, I.W. (2008). Disseminated fungal infection in a renal transplant recipient involving *Macrophominaphaseolina* and *Scytalidiumdimidiatum* : case report and review of taxonomic changes among medically important members of the *Botryosphaeriaceae*. Med Mycol, 46: 285–292.

[3] Crouch, S., Holler, F.A. and Skoog, D.A. (2007). Principles of instrumental analysis. Thomson Brooks Cole, Belmont. ISBN 0-495-01201-7.

[4] Harvey, D. (2000). Spectroscopic methods of analysis. In: Modern analytical chemistry Chapter 10. Mc Graw Hill Publishers, USA, pp. 368–460.

[5] Lopez, M.M., Bertolini, E., Olmos, A., Caruso, P., Gorris, M.T., Llop, P., Penyalver, R. and Cambra, M. (2003). Innovative tools for detection of plant pathogenic viruses and bacteria. Int Microbiol 6: 233–243.

[6] Graeff, S., Link, J. and Claupein, W. (2006a). Identification of powdery mildew (*Erysiphegraminis sp. tritici*) and take-all disease (*Gaeumannomycesgraminis sp. tritici*) in wheat (*Triticumaestivum*L.) by means of leaf reflectance measurements. Central Eur J Biol 1: 275–288.

[7] Graeff, S., Link, J. and Claupein, W. (2006b). Identification of powdery mildew (*Erysiphegraminis* sp. *tritici*) and take-all disease (*Gaeumannomycesgraminis* sp. *tritici*) in wheat (*Triticumaestivum* L.) by means of leaf reflectance measurements. Cent Eur J Biol 1:275–288.

[8] Costa, G., Noferini, M., Fiori, G. and Spinelli, F. (2007). Innovative application of non-destructive techniques for fruit quality and disease diagnosis. ActaHorticulturae 753(1):275–282.

[9] Summy, K.R. and Little, C.R. (2008). Using color infrared imagery to detect Sooty Mold and fungal pathogens of glasshouse-propagated plants. Hortscience 43(5):1485–1491.

[10] Burling, K., Hunsche, M. and Noga, G. (2012). Presymptomatic detection of powdery mildew infection in winter wheat cultivars by laser-induced fluorescence. ApplSpectrosc 66: 1411–1419.

[11] Leufen, G., Noga, G. and Hunsche, M. (2014). Proximal sensing of plant-pathogen interactions in spring barley with three fluorescence techniques. Sensors 14:11135–11152.

[12] Pfeffer, P.E., Bago, B. and Shachar-Hill, Y. (2001). Exploring mycorrhizal function with NMR spectroscopy. New Phytol 150:543–553.

[13] Baranowski, P., Jedryczka, M., Mazurek, W., Babula-Skowronska, D., Siedliska, A. and Kaczmarek, J. (2015). Hyperspectral and thermal imaging of oilseed rape (*Brassica napus*) response to fungal species of the genus *Alternaria*. PLoS ONE, 10, e0012313–e0122913.

[14] Dowell, F.E., Ram, M.S. and Seitz, L.M. (1999). Predicting scab, vomitoxin and ergosterol in single wheat kernels using near-infrared spectroscopy. Cereal Chem 76:573–576.

[15] Pettersson, H. and Aberg, L. (2003). Near infrared spectroscopy for determination of mycotoxins in cereals. Food Control 14:229–232.

[16] Erukhimovitch, V., Tsror, L., Hazanovsky, M., Talyshinsky, M., Mukmanov, I., Souprun, Y. and Huleihel, M. (2005). Identification of fungal phyto-pathogens by Fourier-transform infrared (FTIR) microscopy. J Agric Technol, 1:145–152.

[17] Huang, J. F. and Apan, A. (2006). Detection of Sclerotinia rot disease on celery using hyperspectral data and partial least squares regression. J Spat Sci, 51(2):129–142.

[18] Ramanujam, N., Mitchell, M. F., Mahadevan, A., Thomsen, S., Silva, E. and Richards-Kortum, R. (1994). Luorescence spectroscopy: a diagnostic tool for cervical intraepithelial neoplasia. GynecolOncol, 52(1):31–38.

[19] Belasque, L., Gasparoto, M. C. G. and Marcassa, L. G. (2008). Detection of mechanical and disease stresses in citrus plants by fluorescence spectroscopy. Appl Opt 47(11): 1922–1926.

[20] Maquelin, K., Kirschner, C., Choo-Smith, L.P., Ngo-Thi, N.A., Vreewijk, V., Stammler, M., Endtz, H. P., Bruining, H.A., Naumann, D. and Puppels, G.J. (2003). Prospective study of the performance of vibrational spectroscopies for rapid identification of bacterial and fungal pathogens recovered from blood cultures, J. Clin. Microbiol. 41: 324–329.

[21] Beattie, S. H., Holt, C., Hirst, D. and Williams, A. G. (1998). Discrimination among *Bacillus cereus*, *B. mycoides* and *B. thuringiensis* and some other species of the genus Bacillus by Fourier transform infrared spectroscopy, FEMS Microbiol. Lett. 164: 201–203.

[22] Kummerle, M., Scher, S. and Seiler, H. (1998). Rapid and reliable identification of food borne yeasts by Fourier-transform infrared spectroscopy. Appl. Environ. Microbiol. 64: 2207–2214.

[23] Erukhimovitch, V., Tsror, L., Hazanovsky, M. and Huleihel, M. (2010). Direct identification of potato's fungal phyto-pathogens by Fourier-transform infrared (FTIR) microscopy. Spectroscopy, 24: 609–619.

[24] Salman, A., Tsror, L., Pomerantz, A., Moreh, R., Mordechai, S., and Huleihel. M. (2010). FTIR spectroscopy for detection and identifcation of fungal phytopathogenes. Spectrosc 24(3–4): 261–267.

[25] Salman. A., Shufan, E., Lapidot, I., Pomerantz, A., Huleihel, M., Tsror, L., Moreh, R., and Mordechai, S. (2012). Identifcation of fungal phytopathogens using Fourier transform infrared-attenuated total refection spectroscopy and advanced statistical methods. J Biomed Opt 17(1):017002.

[26] Skolik, P., McAinsh, M. R. and Martin, F. L. (2019). ATR-FTIR spectroscopy non-destructively detects damage-induced sour rot infection in whole tomato fruit. Planta (2019) 249: 925–939.

[27] Almeida, M. R., Alves, R. S., Nascimbem, L. B., Stephani, R., Poppi, R. J. and de Oliveira, L. F. (2010). Anal. Bioanal. Chem. 397: 2693–2701.

[28] Farber, C. and Kurouski, D. (2018). Detection and identification of plant pathogens on maize kernels with a hand-held Raman spectrometer, Anal. Chem. 90: 3009.

[29] Jeanmaire, D. J. and Van Duyne, R. (1977). Surface Raman spectro electrochemistry: Part I. Heterocyclic, aromatic, and aliphatic amines adsorbed on the anodized silver electrode, J. Electroanal. Chem. Interfac. Electrochem. 84: 1.

[30] Pan, T. T., Sun, D.W., Pu, H. and Wei, Q. (2018). Simple approach for the rapid detection of alternariol in pear fruit by surface-enhanced Raman scattering

with pyridine modified silver nanoparticles, J. Agric. Food Chem. 66: 2180.

[31] Sivagnanam, K., Komatsu, E., Rampitsch, C., Perreault, H. and Gräfenhan T. (2017). Rapid screening of *Alternaria* mycotoxins using MALDI-TOF mass spectrometry. J Sci Food Agric. 97(1):357–361.

[32] Brun, S., Madrid, H., Gerrits, Van Den Ende, B., Andersen, B., Marinach-Patrice, C., Mazier, D.(2013). Multilocus phylogeny and MALDI-TOF analysis of the plant pathogenic species *Alternariadauci* and relatives. Fungal Biol. 117(1):32–40.

[33] Chowdappa, P., Lakshmi, M. J. and Madhura, S. (2013). Matrix assisted laser desorption ionization-time of fight (MALDI-TOF) mass spectrometry for identifcation of plant pathogenic *Alternaria* species. Phytoparasitica. 41 (2):169–179.

[34] Chang, S., Porto, Carneiro-Leao, M., Ferreira de Oliveira, B., Souza-Motta, C., Lima, N., Santos, C. (2016). Polyphasic approach including MALDI-TOF MS/MS analysis for identifcation and characterisation of *Fusarium verticillioides* in Brazilian corn kernels. Toxins (Basel).;8(3):5.

[35] Marchetti-Deschmann, M., Winkler, W., Dong, H. J., Lohninger, H., Kubicek, C. P. and Allmaier G. (2012). Using spores for *Fusarium* spp. classifcation by MALDIbased intact cell/ spore mass spectrometry. Food Technol Biotech. 50(3):334–342.

[36] Chalupova, J., Sedlarova, M., Helmel, M., Rehulka, P., Marchetti-Deschmann, M., Allmaier, G., et al. (2012). MALDI-based intact spore mass spectrometry of downy and powdery mildews. J Mass Spectrom. 47(8):978–986.

[37] Beinhauer, J., Lenobel, R., Loginov, D., Chamrad, I., Rehulka, P., Sedlarova, M., et al. (2016). Identifcation of *Bremialactucae* and *Oidiumneolycopersici* proteins extracted for intact spore MALDI mass spectrometric biotyping. Electrophoresis. 37:2940–2952.

[38] Dong, H., Marchetti-Deschmann, M. and Allmaier, G. (2014). Characterization of on-target generated tryptic peptides from *Giberellazeae* conidia spore proteins by means of matrix-assisted laser desorption/ ionization mass spectrometry. Mol Cell Probes.28(2–3):91–98.

[39] Freimoser, F. M., Hilber-Bodmer, M., Brunisholz, R. and Drissner, D. (2016). Direct identification of Monilinia brown rot fungi on infected fruits by matrix-assisted laser desorption/ ionization (MALDI) mass spectrometry. Chem Biol Technol Agricult.3:7.

[40] Neuhof, T., Dieckmann, R., Druzhinina, I. S., Kubicek, C. P., Nakari-Setala, T., and vonDöhren H. (2007). Direct identifcation of hydrophobins and their processing in Trichoderma using intact-cell MALDI-TOF MS. Febs J. 274(3):841–852.

[41] Drissner, D. and Freimoser, F. M. (2017). MALDI-TOF mass spectroscopy of yeasts and flamentous fungi for research and diagnostics in the agricultural value chain. Chem. Biol. Technol. Agric. 4:13.

[42] Sankaran, S., Mishra, A., Ehsani, R. and Davis, C. (2010). A review of advance techniques for detecting plant diseases. Comput. Electron. Agric. 72 (1), 1–13.

[43] EwisOmran, E. (2016). Early sensing of peanut leaf spot using spectroscopy and thermal imaging. Arch. Agron. Soil Sci. https://doi.org/ 10.1080/03650340.2016.1247952.

[44] Martinelli, F., Scalenghe, R., Davino, S., Panno, S. and Scuderi, G. (2015a). Advanced methods of plant

disease detection. A review. Agronomy for Sustainable Development, Springer Verlag/EDP Sciences/INRA, 35 (1): 1–25.

[45] Martinelli, F., Scalenghe, R., Davino, S., Panno, S., Giuseppe Scuderi, G., Ruisi, P., Villa, P., Stroppiana, D., Boschetti, M., Goulart, L. R., Davis, C. E. and Dandekar, A. M. (2015b). Advanced methods of plant disease detection. A review. Agron. Sustain. Dev. 35:1–25.

[46] Martinelli, F., Scalenghe, R., Davino, S., Panno, S., Scuderi, G., Ruisi, P., Villa, P., Stroppiana, D., MircoBoschtti, L., Goulart, R., Davis, C. E. and Dandekar, A. M. (2015c). Advanced methods of plant disease detection. Agron. Sustain. Dev. 35 (1), 1–25.

[47] Ray, S. S., Jain, N., Arora, R. K., Chavan, S. and Panigrahy, S. (2011). Utility of hyperspectral data for potato late blight disease detection. J. Indian Soc. Remote Sens. 39 (2): 161.

[48] Zhang, J. C., Pu, R. L., Wang, J. H., Huang,W. J., Yuan, L. L. and Ju-Hua (2012a). Detecting powdery mildew of winter wheat using leaf level hyper spectral measurements. Comput. Electron. Agric. 85: 13–23.

[49] Zhang J-C, Pu R, Wang J, Huang W, Yuan L, Luo J. (2012b). Detecting powdery mildew of winter wheat using leaf level hyperspectral measurements. Comput Electron Agric.85:13–23.

[50] Romer, C., Burling, K., Hunsche, M., Rumpf, T., Noga, G. and Plumer, L. (2011). Robust fitting of fluorescence spectra for pre-symptomatic wheat leaf rust detection with Support Vector Machines. Comput. Electron. Agric. 79 (2), 180–188.

[51] Jacquemoud, S. and Ustin, S. L. (2001). Leaf optical properties: a state of the art. In: ProcIntSympPhysMeas Sign Rem Sens, pp. 223–232.

[52] Bauriegel, E., Giebel, A., Geyer, M., Schmidt, U. and Herppich, W. B. (2011). Early detection of Fusarium infection in wheat using hyper-spectral imaging. Comput Electron Agric 75: 304–303.

[53] Mahlein, A. K., Oerke, E. C., Steiner, U. and Dehne, H. W. (2012). Recent advances in sensing plant diseases for precision crop protection. Eur J Plant Pathol 133:197–209.

[54] Mahlein, A. K., Rumpf, T., Welke, P., Dehne, H. W., Plümer, L., Steiner, U. and Oerke, E. C. (2013). Development of spectral indices for detecting and identifying plant diseases. Remote Sens Environ 128:21–30.

[55] Reynolds, G. J., Windels, C. E., MacRae, I. V. and Laguette, S. (2012). Remote sensing for assessing Rhizoctonia crown and root rot severity in sugar beet. Plant Dis 96:497–505.

[56] Huang, W., Lamb, D. W., Niu, Z., Zhang, Y., Liu, L. and Wang, J. (2007). Identification of yellow rust in wheat using in-situ spectral reflectance measurements and airborne hyperspectral imaging. Precision Agric 8:187–197.

[57] Singh, A., Ganapathysubramanian, B., Singh, A. K. and Sarkar, S. (2016). Machine learning for high-throughput stress phenotyping in plants. Trends Plant Sci.; 21:110–24.

[58] Mahlein, A. K. (2016). Plant disease detection by imaging sensors–parallels and specific demands for precision agriculture and plant phenotyping. Plant Dis.; 100:241–51.

[59] Fong AY, Wachman E. (2008) Hyperspectral imaging for the life sciences. Biophotonics Int. 15:38.

[60] Vegetation analysis: using vegetation indices in ENVI [Internet].

Exelis VIS [cited 2016 Jan 18]. http://www.exelisvis.com/Learn/WhitepapersDetail/TabId/802/ArtMID/2627/ArticleID/13742/Vegetation-Analysis-Using-Vegetation-Indices-in-ENVI.aspx.

[61] Genc H, Genc L, Turhan H, Smith SE, Nation JL. (2008). Vegetation indices as indicators of damage by the sunn pest (Hemiptera: Scutelleridae) to field grown wheat. Afr J Biotechnol; 7. http://www.ajol.info/index.php/ajb/article/view/58347.

[62] Sims DA, Gamon JA. (2002). Relationships between leaf pigment content and spectral reflectance across a wide range of species, leaf structures and developmental stages. Remote Sens Environ. 81:337–354.

[63] Peñuelas J, Filella I, Lloret P, MUN˜OZ OZ, Vilajeliu M. (1995a). Reflectance assessment of mite effects on apple trees. Int J Remote Sens. 16:2727–33.

[64] Peñuelas J, Baret F, Filella I. (1995b). Semiempirical indices to assess carotenoids/ chlorophyll a ratio from leaf spectral reflectance. Photosynthetica. 31:221–230.

[65] Ashourloo, D., Mobasheri, M. R. and Huete, A. (2014). Developing two spectral disease indices for detection of wheat leaf rust (*Pucciniatriticina*). Remote Sens; 6:4723–40.

[66] Rouse Jr., J. W. 1972 [cited 2016 Feb 29]. Monitoring the vernal advancement and retrogradation (green wave effect) of natural vegetation. http://ntrs.nasa.gov/search.jsp?R=19730009607.

[67] Lasaponara, R. and Masini, N. (2007). Detection of archaeological crop marks by using satellite QuickBird multispectral imagery. J ArchaeolSci.;34: 214–21.

[68] Bravo, C., Moshou, D., West, J., McCartney, A. and Ramon, H. (2003). Early disease detection in wheat fields using spectral reflectance. BiosystEng.; 84:137–45.

[69] Sankaran S, Ehsani R, Inch SA, Ploetz RC. (2012). Evaluation of visible-near infrared reflectance spectra of avocado leaves as a non-destructive sensing tool for detection of laurel wilt. Plant Dis. 96:1683–1689.

[70] Rumpf T, Mahlein A-K, Steiner U, Oerke E-C, Dehne H-W, Plümer L. (2010). Early detection and classification of plant diseases with support vector machines based on hyperspectral reflectance. Comput Electron Agric. 74:91–99.

[71] Moshou D, Bravo C, West J, Wahlen S, McCartney A, Ramon H. (2004). Automatic detection of "yellow rust" in wheat using reflectance measurements and neural networks. Comput Electron Agric. 44:173–188.

[72] Yuan L, Huang Y, Loraamm RW, Nie C, Wang J, Zhang J. 2014. Spectral analysis of winter wheat leaves for detection and differentiation of diseases and insects. Field Crops Res.156: 199–207.

[73] Tian Y, Zhang L. (2012). Study on the methods of detecting cucumber downy mildew using hyperspectral imaging technology. Phys Procedia.;33: 743–750.

[74] Behmann, J., Steinrucken, J. and Plumer, L. (2014). Detection of early plant stress responses in hyperspectral images. ISPRS J Photogramm Remote Sens: 93:98–111.

[75] Mohanty, S. P., Hughes, D. and Salathe, M. (2016). Using deep learning for image-based plant disease detection. ArXiv160403169 Cs [Internet]. 2016 [cited 2016 Sep 12]. http://arxiv.org/abs/1604.03169.

[76] Qin, J., Burks, T. F, Ritenour, M. A. and Bonn W. G. (2009). Detection of

citrus canker using hyperspectral reflectance imaging with spectral information divergence. J Food Eng; 93: 183–91.

[77] Yuhas RH, Goetz AF, Boardman JW. (1992). Discrimination among semi-arid landscape endmembers using the spectral angle mapper (SAM) algorithm. In: Summaries of the third annual JPL airborne geoscience workshop [Internet]. Pasadena, CA: JPL Publication; [cited 2015 Nov 3]. p. 147–9. http://ntrs.nasa.gov/archive/nasa/casi.ntrs.nasa.gov/19940012238.pdf.

[78] Mahlein, A. K., Steiner, U., Hillnhutter, C., Dehne, H. W. and Oerke, E. C. (2012). Hyperspectral imaging for small-scale analysis of symptoms caused by different sugar beet diseases. Plant Methods; 8:3.

Chapter 2

Diagnosis of Fungal Plant Pathogens Using Conventional and Molecular Approaches

Monika C. Dayarathne, Amin U. Mridha and Yong Wang

Abstract

Fungi are a large group of eukaryotes found as saprophytes, pathogens or endophytes, which distribute in every corner of our planet. As the main pathogens, fungi can cause 70–80% of total plant diseases, leading to huge crop yield reduction and economic loss. For identification of fungal plant pathogens, mycologists and plant pathologists have mainly gone through two stages, viz. morphological observation and morphology/phylogeny, and the next era might be utilizing DNA barcodes as the tool for rapid identification. This chapter accounts i) the brief history of development for fungal identification tools and main concepts, ii) the importance and confusion of "One fungus, one name" for pathogen identification, iii) more or fewer species that we need in agricultural practice, and iv) the foreground of fungal plant pathogen identification. These will help to solve the practical problems of identification of fungal pathogens in agricultural production.

Keywords: DNA barcode, morphology, phylogeny, plant diseases, rapid identification

1. Introduction

Plant parasitic fungi are a large group of eukaryotic living organisms lack of photosynthetic pigment and chitinous cell wall. It has been estimated around 15,000 species of them cause diseases in plants [1, 2], and annual crop losses exceed 200 billion euros [3–5]. **Figure 1** shows differently infected plants by various fungal pathogens. During occurrence of plant diseases, they produce various types of essential elements to complete their life cycle [6]. Most of the plants are attacked by one species or several phytopathogenic fungi but also the individual species of fungi can parasitize one or many different kinds of plants [7, 8].

In the pre-molecular era, the detection of fungal pathogens was mostly depending on microscopic, morphological and cultural approaches [9]. The culture-based diagnosis is time consuming and impractical when rapid results are required. With the advancement of molecular methods, detection and identification of phytopathogenic fungi have sped up and become more reliable [4], because of its high degree of specificity to distinguish closely related organisms at different taxonomic levels [10]. Polymerase Chain Reaction (PCR) technologies include multiplex PCR, nested PCR, real-time PCR and reverse transcription (RT)-PCR and DNA barcoding have been recently used as a molecular tool for detection and identification of fungal pathogens [11].

IntechOpen

Figure 1.
A. Leaf spot disease of Houttuynia cordata, B. Downy mildew of Cucumis sativus, C. Peach brown rot of Amygdalus persica, D. Rust disease of Prunus salicina, E-F. Brown rot of Cerasus pseudocerasus leaves and fruit.

The rapid identification of fungal disease is an effective management practice and may help control and prevent their spread and progress successfully. Phylogenetic analyses have been employed for rapid identification of different kinds of fungi. However, the accuracy and reliability of DNA based methods depended largely on the experience and skill of the person making the diagnosis.

Besides that, few plant pathogenic fungi were sometimes also detected and identified using different types of proteomics approaches [6, 11]. In this chapter, we also discussed the importance and confusion of "One fungus, one name", and its impacts on identification of fungal plant pathogens. Finally, some suggestions were referred to the foreground of molecular identification.

2. The brief history of development for fungal identification tools and main concepts

[12] provided a chronological and systematic assessment of conventional methods of plant pathogen identification [13]. The application of light microscopy in the 1840s, the first evidence of plant disease was reported which was caused by *Phytophthora infestans* [14]. In the mid-nineteenth century, spore characters were accepted widely in classification [15]. In the middle of the twentieth century, different fungal structures were given emphasis in taxonomic systems, and separate scientific names (e.g., *Cercospora* were given for more or less similar fungi growing on different plant genera [16]. The observations of ornamentals of spores through scanning electron microscope (SEM) in the mid-1960s helped in separation of very similar plant pathogens and it also aided in clarifying patterns of conidiogenesis [17]. Then when came to the era of Transmission Electron Microscopy (TEM) which led to the discovery of fundamental differences in the major groups [18]. **Figures 2** and **3** represent the ultrastructural morphology of spongy tissue cells of tea leaves infected by fungal pathogens and control leaves by TEM [19].

During 1960s and 1970s thin-layer chromatography (TLC) and isozyme profiles were used to find out the chromosome numbers [20]. Vegetative compatibility groups (VCGs) were developed and it was found importance in many research studies on pathogenic *Fusarium* spp. [21]. The cluster analysis was performed

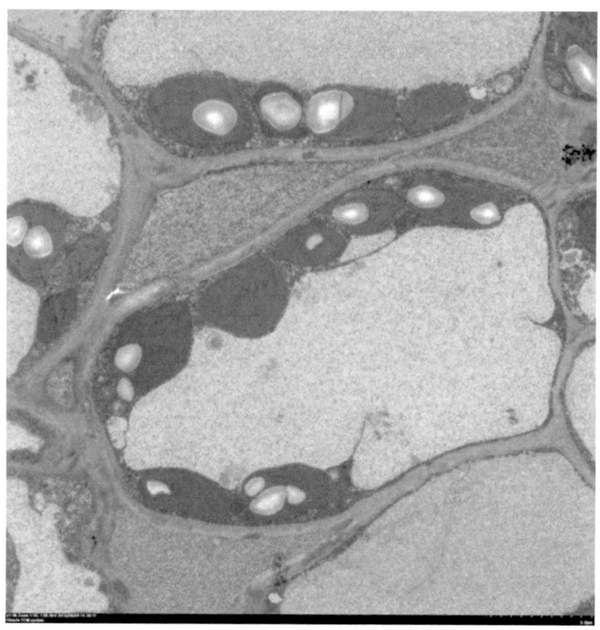

Figure 2.
The healthy spongy tissue of tea leaves, observed by TEM.

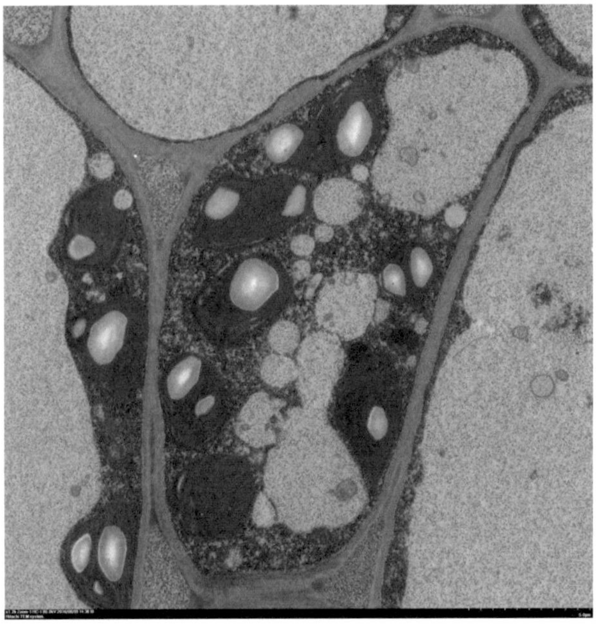

Figure 3.
Exobasidium vexans infects spongy tissue cells of tea leaves, observed by TEM.

after having powerful computers in the 1970s which revealed large numbers of morphological, cultural, and physiological characteristics should be computed and analyzed together. With that situation, DNA-based methodologies moved from occasional to common use [22].

[23] mentioned that the identification of fungi up to generic/species level or the *formae speciales* strains depended on their morphological characteristics and various kinds of reproductive organs. However, varieties or biotypes have to be identified by following pathogenicity, biochemical and immunological properties or nucleotide sequences of the genomic DNA, isozyme analysis, vegetative compatibility group (VCG) analysis and electrophoretic mobility of cell wall proteins etc. The development of enzyme-linked immunosorbent assay (ELISA) and monoclonal antibodies exhibit greater sensitivity and specificity in identifying fungi [23].

The molecular technologies are widely used in identifying plant pathogenic fungi and have been studied by mycologist and plant pathologists throughout the world. Many different types of diagnostic techniques may be used for detection, identification and quantification of fungal pathogens present in the infected above ground and below ground parts of plants and propagating and reproductive organs of different types of plants [23]. The nucleotide sequences of the pathogen DNA have become the preferred ones, because of their greater speed, specificity, sensitivity, reliability, and reproducibility of the results obtained, following the development of PCR [23]. [24] mentioned that the researchers over the last few years devoted their affords to develop the methods for detecting and identifying plant pathogens based on DNA/RNA probe technologies and PCR amplification such as [25] developed techniques for the rapid detection of plant pathogens; [26] used PCR for identifying plant pathogens; [27] used the modern assays for identification, detection and quantification of plant pathogenic fungi: and impacts of molecular diagnostic technologies on plant disease management was evaluated by [28]. The RT-PCR advances are helping the accurate detection and quantification of plant pathogens quickly and now being used routinely in most of the aspects of plant pathology.

In all molecular technology, DNA technology is most important in recovering from living cultures but is also useful to revise major groups of obligate fungi that cannot be cultivated, such as the powdery mildews [29], rusts and smuts [30]. Whole-genome sequence analyses indicated that the millions of dried fungal specimens preserved in different collection centers could hold great promise for understanding the evolution of many major fungal pathogens and their associated diseases and epidemics over time [31, 32]. [33] described a large number of various important common leaf diseases (from 2004 to 2019) caused by fungal plant pathogens with their symptoms and references of publications. [34] mentioned that the PCR and flow cytometry may be used in the genetic recognition of existing pathogens and the identification of emergent ones. The minute quantities of DNA in plant pathogens may be detected because of sensitiveness of DNA-based PCR technologies [35]. Further, genetic investigations could detect sources of pathogen and host resistance in diseases such as powdery mildew. [4] mentioned the different molecular diagnostics techniques (**Table 1**) used by many researchers throughout the world for the identification of phytopathogenic fungi with their advances and disadvantages.

Accurate identification and diagnosis of plant pathogens with reliable technologies and methods are needed to control them for sustainable plant diseases management [84] as well as prevention of the spread of invasive pathogens [85]. [86] published their works on fungal protocols and the primers for the ITS were first introduced, and it is still valid and widely used [32] in identification of plant pathogens. [9] reported that species identification was frequently difficult because fungi are a large and diverse assemblage of eukaryotes and have complex and

Molecular method	**Reference**
Conventional PCR	[36, 37]
Nested PCR	[38–40]
Multiplex PCR	[41]
Reverse transcriptase (RT) PCR	[10, 42]
Real-time PCR (Q PCR)	[43–45]
Serial analysis of gene expression (SAGE)	[46, 47]
DNA barcoding	[32, 48–51]
DNA/RNA probe-based methods	[24]
Northern blotting	[52–54]
In situ hybridization	[55–57]
FISH	[58, 59]
Post amplification techniques	[60–62]
Macroarray	[62–64]
The isothermal amplification-based methods	[58, 65, 66]
Loop-mediated isothermal amplification (LAMP)	[66–71]
Nucleic acid sequence-based amplification (NASBA):	[68, 72–75]
RNA interference methods (RNAi)	[76–79]
RNA-Seq-based next-generation sequencing methods	[46, 80–83]

Table 1.
PCR-based molecular methods for the detection of fungi.

poorly understood life cycles [87]. They have mentioned that molecular (DNA sequence) data as an essential tool for the identification of plant pathogenic fungi by the nuclear ribosomal internal transcribed spacer (ITS) region. The barcode gene for the fungi could be used to identify a wide range of plant-pathogenic fungi [9]. Protein-coding genes [glyceraldehyde-3-phosphate dehydrogenase (GAPDH), beta-tubulin (*tub2*) gene, translation elongation factor 1-alpha (tef1), actin (act), and histone H3 (his3)], generally prove a valuable supplement to ribosomal genes at the species level. More conserved gene regions such as large subunit (LSU), small subunit (SSU), and RNA polymerase II (RPB2) gene provide a better discrimination at the generic and/or family level [88–90]. The Q-bank fungal database contains DNA barcodes supplemented by morphological, phenotypical, and ecological data for more than 725 species of relevance to phytopathology. The database continues to be actively expanded, and parties interested in participating or contributing can contact its curators (http://www.q-bank.eu). The molecular identification of fungi (http://unite.ut.ee), is available on the basis of the results from a total of 31,954 changes incorporated and made available through the UNITE database, standalone FASTA files of sequence data for local BLAST searches and also use in the next-generation sequencing analysis platforms QIIME and mother [9]. The results were incorporated in UNITE made available publicly (http://unite.ut.ee/repository.php) e.g., local sequence similarity searches and sequence processing pipelines such as QIIME [91, 92], mothur [93], SCATA (http://scata.mykopat.slu.se/), CREST [94], and other downstream applications. UNITE also serves as one of the data providers for BLAST [95] searches in the EUBOLD fungal barcoding database (http://www.cbs.knaw.nl/eubold/). The maximum parsimony, maximum likelihood, and/or Bayesian inference are currently practiced to identify in many genera of phytopathogenic fungi [12]. Interestingly, improvement in molecular techniques has begun to allow a rapid alternative rDNA sequencing to whole genome sequencing [96]. LAMP of DNA is a newer molecular technology for affordable, specific, highly sensitive, and rapid diagnostic testing of pathogens in both laboratory and field conditions [97], and subsequently been optimized for portable instruments in field. Recently several protocols for a rapid detection of woody pathogens, such as *Ceratocystis platani, Fusarium circinatum, F. euwallaceae, Xylella fraxineus,* and *Phytophthora ramorum,* have been established [98–100].

Additionally, the application of proteomics such as two-dimensional gel electrophoresis (2-DE) and mass spectrometry (MS) is used to characterize cellular and extracellular virulence and pathogenicity factors produced by pathogens as well as to identify changes in protein levels in plant hosts upon infection by pathogenic organisms and symbiotic counterparts [101]. Many of the techniques used in proteomics, in particular the 2-DE method was developed two decades before the term proteomics was coined [102, 103]. Two-dimensional gel electrophoresis (2-DE) have been carried out to study the proteome of phytopathogenic fungi, mainly due to the difficulty of obtaining fungal protein extracts and/or the lack of available fungal protein databases [6]. In the last few years proteomic, in conjunction with genomic, has become one of the most relevant techniques for studying phytopathogenic fungi. Currently, the complete genome of over four hundred different species has been sequenced and this number is still increasing. With the availability of genome information for more and more species and the advancement in mass spectrometry technologies, proteomics has come into true since 1990s [104]. The advent of proteomics has allowed researchers to identify a broad spectrum of proteins in living systems.

Almost a little earlier, [86, 105–107] used immunological techniques with fungal plant pathogens-aspects of antigens, antibodies and assays for diagnosis. To speed up the identification of plant pathogens and allow their identification in field, a number of serological methods have been developed, mainly based upon the enzyme-linked immunosorbent assay (ELISA). These methods are used to detect pathogens using a monoclonal antibody labeled with fluorescent compounds [108, 109]. Lateral flow devices (LFDs) are a simple paper-based dipstick assay able to detect and identify the causal agents of disease [110, 111]. [112] also described the methods using allozyme and isozyme markers to rapidly differentiate intersterility groups of *Heterobasidion annosum* [113], *Phytophthora cinnamomi,* and *Seiridium* sp. isolates [114].

3. The importance and confusions of "One fungus, one name" concept in plant pathogen identification

3.1 "One fungus, one name" concept

Scientific names (Latin binomials) are an integral part for communicating details about fungi causing plant diseases. Assembled knowledge on fungal pathogens *viz* biology, distribution, ecology, host range, control measures and the risks are accessible through these names [12]. In naming new fungal species mycologists are governed by the ICN and, more specifically by the *International Code for Nomenclature of plants, algae, and fungi* [115]. The Code provides a platform to abolish any bias or taxonomic confusion where multiple names are used for the same species [116]. '1 Fungus = 1 Name' was a meeting organized by CBS in the Netherlands that resulted in the 'Amsterdam Declaration' signed by some 80 participants [117], that strictly proposed the move to a unified nomenclature. Each fungal species should have one accurate name which is nomenclaturally accepted in a particular classification. Any of the previously used names for a particular species should be considered as a synonym with the oldest epithet taking priority over any younger name. If there is a wish/desire/movement to use a widely known younger name, then such usage must be in accordance with Art. 57.2 of the Code and its adoption should be accepted by the Nomenclature Committee for Fungi (NCF). Nevertheless, application of "one fungus one name" (1F1N) is in its infancy in mycology because most of the fungi are commonly known in only their sexual or asexual morph [116].

3.2 Significance of 1F1N concept in plant pathogen identification

Pleomorphism (having diverse fungal propagules) can be seen in many pathogenic fungi especially in Ascomycetes and in basidiomycetous rust fungi [118, 119]. Until the early 2000s, fungi were primarily classified on the basis of their sporing structures, and separate names were given to the sexual structures (formerly called the teleomorph) and asexual structures (formerly called the anamorph or if there are several asexual morphs, synanamorphs) even where the relationship between different morphs was proved by the culture of single spores [119]. For example, *Calonectria* with *Cylindrocladium* asexual morphs [120], *Chaetosphaeria* with *Menispora* asexual morphs [121], *Cladosporium* with *Venturia* asexual morphs [122], *Gibberella* with *Fusarium* asexual morphs [122], *Ceratocystis* with *Thielaviopsis* asexual morphs [123] and *Grosmannia* with *Leptographium* asexual morphs [124].

However, the concept of dual name became controversial to mycology in 21st century especially when a single DNA sequence could be attached to two names; one being the sexual morph name and the other under the asexual morph name (e.g. species of *Diaporthe* and *Phomopsis*) [125]. Many people working in fields related to agriculture/horticulture and plant pathology are confused by having to deal with two names for a single pathogen [119]. This can be very important when dealing with fungi of quarantine significance and quarantine regulations linked to import and export requirements. Some countries may list the asexual morph name for an organism, whereas others list the sexual morph name. It is true that the two names refer to the one genetically identical organism, but quarantine officers are not necessarily aware of these details when dealing with constantly changing asexual–sexual morph taxonomy. For instance, identification of an invasive new rust (*Uredo rangelii*) on Myrtaceae in Australia [126]. This raised confusion as to whether or not the much-feared Eucalyptus rust (*Puccinia psidii*), a serious quarantine organism, and a restricted fungus on quarantine lists in countries in which eucalypts are cultivated [127, 128], was identical and had been introduced into Australia. Genetically, these names represent the same fungus or, at least, very closely related fungi causing the same disease, which suggests that they should be treated in a similar fashion when it comes to quarantine decisions. However, the names have not been treated equally and this has caused substantial complications relating to the treatment of the new *P. psidii sensu lato* invasion in Australia [126].

Dual nomenclature also conflicts with biological philosophy; a type is the type of a single organism that can have only one legitimate name [129]. The concept of permitting separate names for asexual morph of fungi with a pleomorphic life-cycle has been also an issue for mycologists to collect and describe new fungal species, mostly with one morph [116]. Therefore, depending on the accepted recommendations of 1F1N concept, all legitimate fungal names are now treated equally for the purposes of establishing priority. Asexual morph genera compete with sexual morph genera based on priority. For example, the asexual genera names *Alternaria* (1817) takes precedence over the sexual genus name *Lewia* (1986), *Cladosporium* (1816) over *Davidiella* (2003), *Fusarium* (1809) over *Gibberella* (1877), *Phyllosticta* (1818) over *Guignardia* (1892), *Sphaceloma* (1874) over *Elsinoë* (1900), *Trichoderma* (1794) over *Hypocrea* (1825). However, the reverse can also happen where an older sexual genus name takes priority over a younger asexual genus name, e.g., *Diaporthe* (1870) over *Phomopsis* (1905). However, there are exceptions where younger, widely used names get priority over an older name, for example *Hypomyces* (1860) over *Cladobotryum* (1816).

[130] documented five alternatives which can be followed when deciding on a single name for a fungus with a pleomorphic life cycle. These are: 1) strict priority, ignoring names originally typified by asexual morph or sexual morph by considering the priority of both generic names and species epithets [131, 132]; 2) sexual morph priority, with asexual morph species epithets [133]; 3) sexual morph priority without considering earlier asexual morph species epithets [134–136]; 4) teleotypification and 5) single species names but allowing two genera per clade (*Hypomyces*/*Cladobotryum*) [137, 138]. A number of sexual and asexual morph fungal genera have been linked by applying the oldest available name for the lineage (strict priority) in various studies. For example, *Neofusicoccum* was assigned for the clade with unnamed *Botryosphaeria*-like sexual morphs [139] included asexual *Phialophora*-like fungi in the sexual morph genus *Jattaea* [120, 140–142]; *Cylindrocladium* species were included under the older generic name *Calonectria*, and *Phomopsi* species in the older, sexual genus *Diaporthe* [139]. Importantly, 1F1N is important to link asexual morphs of pathogenic fungi to sexual morph-typified generic names, even without ever having seen the sexual morphs (e.g *Teratosphaeria toledana* and *Phaeophleospora toledana*) [119]. Further, this approach is also crucial for the widely emerging whole genome sequencing projects specially to compare species representing single entities with their closest relatives [143]. Such as comparing *Mycosphaerella tritici* (now *Zymoseptoria*) with *Mycosphaerella fijiensis* (now *Pseudocercospora*), is not instructive, as they are just two genera within a family, but not two species of one genus [119].

4. Controversies associated with one fungus one name concept in fungal plant pathogen identification

Although the application of 1F1N has become a reality, determination of which name to use for certain fungal species is somewhat more complex. Also, it is doubtful when accepting the other morph if it has been described elsewhere with a different name especially when lacking molecular data [116]. According to 1F1N mycologists must now select a genus name formerly applied to taxa with either asexual or sexual reproductive modes, that decision often influences the scope of genotypic and phenotypic diversity of a genus, and even its monophyly. [144] showed that many pairs of legitimate asexual-sexual morph names are not homotypic synonyms and merging them may not be justified. Therefore, dual names continue to be available for use following [145] e.g. the name pairs *Aspergillus niveus – Fennellia nivea* and *Aspergillus flavipes – Fennellia flavipes*, were not conspecific in a molecular study by [146].

Another problem arises when pathogenic species have one or more generic names for sexual morph associated with one or more asexual states. The best example is *Aspergillus* species which are mostly opportunistic pathogens. There are 11 sexual generic names associated with this genus; phenotypic variation and genetic divergence within the asexual genera are low but between sexual genera they are high [147]. Applying the asexual name *Aspergillus* to the many sexual genera masks information now conveyed by the sexual genus names. This would lead to taxonomic inconsistency in the Eurotiales because the large *Aspergillus sensu lato* would embrace more genetic divergence than neighboring clades comprised of two or more genera. However, [148] proposed a phylogeny combined aspect to apply one name to one fungal genus in a scientific manner in such a case.

Establishment and full use of the single name concept may take a long time as it is difficult to discard fungal names in publications before 2013, and these materials are still in use. The old name of some species whose name has been changed is still used in many publications [148]. When identifying fungi that cause diseases in humans, animals or plants, it may be difficult to determine which is the correct name because there are different names for these fungi in the literature. It is unlikely that all researchers and workers in agricultural industries, or border protection officers will have a good knowledge and understanding of fungal taxonomy. Acceptance and widespread use of the fungal names that change due to 1F1N will take time. Therefore, in some ways we are trapped in the past and there is difficulty in applying recent knowledge, due to long-standing and traditional rules that define how we name fungi.

5. More or fewer species that we need in agricultural practice

With the advent of "One fungus, one name" times from 2010s, many important fungal genera and species, for example *Gibberella*, *Hypocrea*, *Phomopsis* and *Magnaporthe grisea* causing worldwide rice blast towards the end became the synonyms of *Fusarium*, *Trichoderma*, *Diaporthe* and *Pyricularia grisea* respectively approved by the Nomenclature Committee of the Fungi and the General Committee (Art. 14.13). [149] listed nearly 7,000 generic names for eventual adoption, which made up just less than 50% of the total [24, 38] legitimate generic names) from Index Fungorum/ MycoBank database. For these changes, molecular techniques play an important role in the emergence of this great change, although for the species concept of fungi, we do still not get rid of the cruse of pragmatism. Thus, this has led to a very puzzling phenomenon, viz. on one side oceans of known species walk towards death, but on the other side mycologists spare no effort to 'create' many new species and even many higher-level taxa (genus, family, and order, etc.). *Trichoderma harzianum* as an ubiquitous species in the environment and also effective bio-control agents against the devastating plant diseases, became an aggregate species recognized by [150], using genealogical concordance and recombination analyses confirmed there were two genetically isolated agamospecies and two hypothetical holomorphic species related to *T. harzinanum* species-complex [151], but surprisingly split into at least 14 species based on morphological, ecological, biogeographical and phylogenetic data [152, 153]. For *Alternaria* and allied genera, even the whole Kingdom Fungi, 2013 was destined to go down in history because of "*Alternaria* redefined", up to eighteen old generic name, for example, *Embellisia*, *Nimbya*, *Ulocladium* and *Lewia* turned into the synonyms of *Alternaria*, but in the meantime, 16 new *Alternaria* section were born [154].

Immediately, Hyde and Crous as well as their research groups open the dazzling "re-" doors published in Fungal Diversity, Studies in Mycology, Persoonia, IMA Fungus, Mycosphere. They provided a series of "backbone" trees of fungal genera, family, order or even higher taxonomic level based on DNA sequences from ex-type, epitype and authoritative strains. From 2014 to 2020, "One stop shop: backbones trees for important phytopathogenic genera: I-IV", were published in Fungal Diversity and led by [155] and [156–158] with international co-operations, which provided phylogenetic frameworks of 100 groups or genera of plant pathogenic fungi in the Ascomycota, Basidiomycota, Mucormycotina (Fungi), and Oomycota. Almost at the same time, in Studies in Mycology, a series of "Genera of phytopathogenic fungi: GOPHY1-3", which introduced stable platforms for the taxonomy of 62 phytopathogenic genera, including 5 new genera, 88 new species, 38 new combinations, four new names and 13 typifications of older names [159–161]. For these publications, the important disease information, viz. distribution, hosts and disease symptoms were referred, but without the key pathogenicity test (Koch's postulates) to clarify whether they were real pathogens or not. In spite of this, these contributions still make us get rid of the embarrassment of using morphology as the only approach of pathogen identification and provide primary and secondary DNA barcodes for rapid and accurate recognition. After census of new pathogens report in the international mainstream journals of plant pathology, we discovered that in the latest three years, more than 200 new pathogens and first reports were recorded per year in our planet.

Now more and more mycologists and plant pathologists accepted that fungi causing plant or post-harvest diseases should be identified on the basis of morphology and phylogeny or at least ITS-blast on NCBI database (for example, https://www.apsnet.org/publications/plantdisease). Especially, [162] solemnly declared that the optimal identity thresholds to discriminate filamentous fungi on the species level were 99.6% for ITS and 99.8% for LSU regions using more than 24,000 DNA barcode sequences originated from 12,000 ex-type strains. Even so, for important plant pathogenic fungal group (*Alternaria*, *Botrysphaeria*, *Colletotrichum* and *Diaporthe*), if only sequences of ITS or LSU region, the result will be considered rash and superficial. We have to admit that for identification of fungal pathogens, the agricultural practitioners welcomed fewer and simpler, but mycologists always looked ahead into the future and back into the past to creation or elimination.

6. The foreground of fungal plant pathogen identification

Accurate identification of pathogens must be the first step of plant pathology. Linnaeus published "Species Plantarum" in 1753 and then "Systema Naturae" (10th edition) in 1758 for planting naming with binomial nomenclature, which were continued in Kingdom Fungi. A dual system of fungal nomenclature for asexual fungi was promulgated by [15], at one time, which played an important role in the identification of plant pathogenic fungi but came to the end in 2013. [147] compared the distinction between theoretical and operational species concepts, and pointed that PSR (Phylogenetic Species Recognition) by genealogical concordance was well suited to fungi and developed and adopted at an increasing rate [163]. DNA barcode, as a relative short specific DNA sequence was able to utilize in taxonomic practice referring to OTUs (Operational Taxonomic Units), which was comprehensively discussed by [164]. Urgently [165, 166] even attempted to propose DNA sequences without vouchered specimens to serve as types for fungal taxon names, but was unfortunately rejected by Nomenclature Committee for Fungi and International Mycological Congress (IMC 11) [167]. Almost at the same time, [168] further pointed out ASVs (Amplicon Sequence Variants) could replace OTUs as the standard units by high-throughput marker-gene sequencing data analysis.

The rapid development about identification approach of fungi has entered a dazzling but seemingly at a loss stage in plant pathology and other related practical or applied scientific fields. Although this, we have to admit the reality or the status quo is existing mycological research networks, especially e-books or publications do really facilitate the rapid development of DNA identification and information sharing. We can even update our knowledge in almost days and more comprehensive. It can also be understood in this way, viz. easier to make mistakes but also correct them. Although [167] fully expounded the deficiencies of Hawksworth's proposals, for identification of plant pathogenic fungi, we believe that accuracy sometimes gives way to quickness. Thus, DNA identification is competent to become a core or sole approach for fungal pathogens.

For plant pathologists in consideration of this method, we can quickly start the following two works, i) to make full use of the achievements of taxonomists to all-round confirm or correct the scientific name of old fungal pathogens, like "one fungus, one name" and "backbone trees" of fungal groups, which needs to be simultaneously done by pathologists in different countries of the world, or at least one continent, and 2) to standard the identification parameters of plant pathogenic fungi, for example the barcoding gene markers (only ITS or ITS plus a secondary generic marker) for PCR amplification (including forward/reverse primers), sequences threshold (99.6% for ITS or 99.8% for LSU is OK, or adopt the new standard?) and international specialized open database for rapid alignment. Of course, we also should keep pace with mycologists, and update our identification system on time.

Author details

Monika C. Dayarathne[1], Amin U. Mridha[1,2] and Yong Wang[1*]

1 Department of Plant Pathology, Agricultural College, Guizhou University, Guiyang, China

2 Department of Botany, University of Chittagong, Bangladesh

*Address all correspondence to: yongwangbis@aliyun.com

IntechOpen

© 2020 The Author(s). Licensee IntechOpen. This chapter is distributed under the terms of the Creative Commons Attribution License (http://creativecommons.org/licenses/by/3.0), which permits unrestricted use, distribution, and reproduction in any medium, provided the original work is properly cited.

References

[1] Hawksworth DL, Kirk PM, Sutton BC. Pegler DN. Ainsworth and Bisby's Dictionary of the Fungi, International Mycological Institute, Surrey, UK, 2001.

[2] Hawksworth DL. The fungal dimension of biodiversity: magnitude, significance, and conservation, Mycol. Res. 1991;95(6):641-655.

[3] Arora DK, Bridge PD, Bhatnagar D. Fungal Biotechnology in Agricultural, Food and Environmental Applications, Marcel Dekker, New York, NY, USA, 2004.

[4] Aslam S, Tahir A, Aslam MF, Alam MW, Shedayi AA, Sadia S. Recent advances in molecular techniques for the identification of phytopathogenic fungi – a mini review. J. Plant Interact. 2017;12:493-504, DOI: 10.1080/17429145.2017.1397205.

[5] Ristaino JB. Tracking historic migrations of the Irish potato famine pathogen, *Phytophthora* infestans. Microbes and Infection 2002;4:1369-1377. DOI:10.1016/S1286-4579(02)00010-2.

[6] Fernández-Acero FJ, Carbú M, Garrido C, Vallejo I. Cantoral JM. Proteomic Advances in Phytopathogenic Fungi. Current Proteomics, 2007;4:79-88.

[7] Agrios. 5th ed., Plant Pathology WSU, OSU, U of I, 2005, Pacific Northwest Plant Disease. Handbook Moore, Randy1998, Botany, 2005. http:/www.mycolog.com/fifthloc.html.

[8] Tao WC, Zhang W, Yan JY, Hyde KD, McKenzie EH, Li XH, Wang Y. A new Alternaria species from grapevine in China. Mycol. Prog. 2014;13(4):999.

[9] Nilsson RH, Hyde KD, Pawłowska J, Ryberg M, Tedersoo L, Aas AB, Alias SA, Alves A, Anderson CL, Antonelli A, Arnold AE. Improving ITS sequence data for identification of plant pathogenic fungi. Fungal Divers. 2014;67:11-19. DOI 10.1007/s13225-014-0291-8.

[10] Capote N, Pastrana AM, Aguado A, Sánchez-Torres P. Molecular tools for detection of plant pathogenic fungi and fungicide resistance. In: Cumagun CJ (Ed.), Plant Pathology, InTech, Rijeka, Croatia; 2012. 362 p.

[11] González FR, Jorrín Novo JV. Proteomics of fungal plant pathogens: the case of *Botrytis cinerea*. Current Research, Technology and Education Topics in Applied Microbiology and Microbial Biotechnology, A. Méndez-Vilas (Ed.), 2010;205-216.

[12] Crous PW, Hawksworth DL, Wingfield MJ. Identifying and Naming Plant pathogenic Fungi: Past, present and Future. Annu. Rev. Phytopathol. 2015;53:247-267

[13] Crous PW, Slippers B, Wingfield MJ, Rheeder J, Marasas WF, Philips AJ, Alves A, Burgess T, Barber P, Groenewald JZ. Phylogenetic lineages in the Botryosphaeriaceae. Studies in mycology. 2006;55:235-253. DOI:10.3114/sim.55.1.235.

[14] Berkeley MJ. Observations, botanical and physiological, on the potato murrain. J. R. Hortic. Soc. 1846;1:9-34

[15] Saccardo PA. De Diagnostica et nomenclatura mycologica, Admonita quaedam. Ann. Mycol. 1904;2:195-198. [English translation by Clements, F.E. J. Mycology. 1904;109-112.

[16] Chupp C. A Monograph of the Fungus Genus *Cercospora*. Ithaca, NY, 1954.

[17] Cole GT, Samson RA. *Patterns of Development in Conidial Fungi*. London: Pitman Publ. 1979.

[18] von Arx JA, van der Walt JP, Liebenberg NVDM. The classification of *Taphrina* and other fungi with yeast-like cultural states. Mycologia, 1982;74:285-296.

[19] Zhao XZ, Wang Y, Ren YF, Li DX, Chen Z. The morphology observation of infection process for the pathogen *Exobasidium vexans* of tea blister blight against tea leaf. Chinese Agri Sci Bull. 2020;34:117-122

[20] Wieloch W. Chromosome visualisation in filamentous fungi. J. Microbiol. Meth. 2006;67:1-8

[21] Leslie JF, Summerell BA, eds. The *Fusarium* Laboratory Manual. Ames, IA: Blackwell Publishers 2006.

[22] Schloss PD, Westcott SL, Ryabin T, et al. Introducing mothur: open-source, platform-independent, community-supported software for describing and comparing microbial communities. Appl. Environ. Microbiol. 2009;75:7537-7541.

[23] Narayanasamy, P. Detection of Fungal Pathogens in Plants. Microbial Plant Pathogens-Detection and Disease Diagnosis. 2010;5-199.

[24] McCartney HA, Foster S J, Fraaije BA, Ward E. Molecular diagnostics for fungal plant Pathogens. Pest Manag Sci. 2003;59:129-142 DOI: 10.1002/ps.575.

[25] Duncan JM, Torrance L. Techniques for the rapid detection of plant pathogens, Blackwell Scientific Publications, Oxford .1992.

[26] Ward E. Use of the Polymerase Chain Reaction for identifying plant pathogens, in Ecology of plant pathogens, In: Blakeman JP, Williamson B. (editors) CAB International, Wallingford, UK, 1994; pp. 143-160

[27] Schots A, Dewey FM, Oliver R. (eds), Modern assays for plant pathogenic fungi: Identification, detection and quantification, CAB International, Wallingford, UK. 1994.

[28] Martin RR, James D, Le'vesque CA. Impacts of molecular diagnostic technologies on plant disease management. Annu. Rev. Phytopathol. 2000;38:207-239.

[29] Takamatsu S, Kano Y. PCR primers useful for nucleotide sequencing of rDNA of the powdery mildew fungi. Mycoscience 2001;42:135-139.

[30] Shivas RG, Beasley DR, McTaggart AR. Online identification guides for Australian smut fungi (*Ustilaginomycotina*) and rust fungi (*Pucciniales*). IMA Fungus, 2014;5:195-202.

[31] Gibbons A. On the trail of ancient killers. Science. 2013;340:1278-1282

[32] Schoch CL, Seifert KA, Huhndorf S, Robert V, Spouge JL, et al. Nuclear ribosomal internal transcribed spacer (ITS) region as a universal DNA barcode marker for *Fungi*. Proceedings of the Nat. Acad. Sci. 2012;109:6241-6246.

[33] Jaina A, Sarsaiya S, Wua Q, Lua Y, Shi J. A review of plant leaf fungal diseases and its environment speciation. Bioengineered. 2019;10(1):409-424. https://doi.org/10.1080/21655979.2019.1649520

[34] Sarsaiya S, Jia Q, Fan X, et al. First report of leaf black circular spots on Dendrobium nobile caused by *Trichoderma longibrachiatum* in Guizhou Province, China. Plant Dis. 2019. DOI:10.1094/pdis-03-19-0672-pdn

[35] Lee SB, Taylor JW. Isolation of DNA from fungal mycelia and single

spores, in PCR protocols. A guide to methods and applications, In: Innis MA, Gelfand DH, Sninsky JJ and White TJ, (editors) Academic Press, San Diego, USA; 1990. pp. 282-287

[36] Fang Y, Ramasamy RP. Current and prospective methods for plant disease detection. Biosensors. 2015;5(3):537-561.

[37] Mancini V, Murolo S, Romanazzi G. Diagnostic methods for detecting fungal pathogens on vegetable seeds. Plant Pathol. 2016;65:691-703.

[38] Bhat R, Browne G. Specific detection of Phytophthora cactorum in diseased strawberry plants using nested polymerase chain reaction. Plant Pathol. 2010;59(1):121-129.

[39] Tsai HL, Huang LC, Ann PJ, Liou RF. Detection of orchid Phytophthora disease by nested PCR. Bot Stud. 2006;47(4).

[40] Yeo SF, Wong B. Current status of nonculture methods for diagnosis of invasive fungal infections. Clin. Microbiol. Rev. 2002;35(3):465-484

[41] Cho HJ, Hong SW, Kim HJ, Kwak YS. Development of a multiplex PCR method to detect fungal pathogens for quarantine on exported cacti. Plant Pathol. J. 2016;32(1):53-54.

[42] Brown NA, Bass C, Baldwin TK, Chen H, Massot F, Carion PW, Urban M, Van De Meene AM, Hammond-Kosack KE. Characterisation of the *Fusarium graminearum*-wheat floral interaction. J. Pathogens, 2011;9. DOI:10.4061/2011/626345

[43] Badali H, Nabili M. Molecular tools in medical mycology; where we are! Jundishapur J. Microbiol. 2012;6(1):1-3.

[44] Black J. Quantitative real-time polymerase chain reaction (qPCR) of filamentous fungi in carpet. Research Triangle Park, NC: RTI Press; 2009.

[45] Dasmahapatra K, Mallet J. DNA barcodes: recent successes and future prospects. Heredity. 2006;97(4):254-255.

[46] Dawei W, Peng Y. Assessing the impact of dominant sequencing base gene expression profiling techniques (SGEPTs) on phytopathogenic fungi. Chiang Mai J. Sci. 2014; 41:922-9444.

[47] Velculescu VE, Zhang L, Vogelstein B, Kinzler KW. Serial analysis of gene expression. Science. 1995;270(5235):484-487.

[48] Krishnamurthy PK, Francis RA. A critical review on the utility of DNA barcoding in biodiversity conservation. Biodiversity and Conservation. 2012;21 (8):1901-1919.

[49] Li DZ, Gao LM, Li HT, Wang H, Ge XJ, Liu JQ, Chen ZD, Zhou SL, Chen SL, Yang JB. Comparative analysis of a large dataset indicates that internal transcribed spacer (ITS) should be incorporated into the core barcode for seed plants. Proc. Natl. Acad. Sci. 2011;108 (49):19641-19646..

[50] Monteiro F, Romeiras MM, Figueiredo A, Sebastiana M, Baldé A, Catarino L, Batista D. Tracking cashew economically important diseases in the West African region using metagenomics. Front. Plant Sci. 2015;6:482-483.

[51] Roe AD, Rice AV, Bromilow SE, Cooke JE, Sperling FA. Multilocus species identification and fungal DNA barcoding: insights from blue stain fungal symbionts of the mountain pine beetle. Mol Ecol. Resour. 2010;10(6):946-959.

[52] Berg Jeremy M. Biochemistry. 6th ed. New York: W. H. Freeman and Company; 2007.

[53] Kim SW, Li Z, Moore PS, Monaghan AP, Chang Y, Nichols M, John B. A sensitive non-radioactive northern blot method to detect small RNAs. Nucleic Acids Res.2010;38:e98..

[54] Qi M, Yang Y. Quantification of *Magnaporthe grisea* during infection of rice plants using real-time polymerase chain reaction and northern blot/ phosphoimaging analyses. Phytopath. 2002;92 (8):870-876.

[55] Ellison MA, McMahon MB, Bonde MR, Palmer CL, Luster DG. In situ hybridization for the detection of rust fungi in paraffin embedded plant tissue sections. Plant Methods. 2016;12(1):37

[56] Jensen E. Technical review: in situ hybridization. Anatomical Records 2014; 297(8):1349-1353.

[57] Qian X, Lloyd RV. Recent developments in signal amplification methods for in situ hybridization. Diagn. Mol. Pathol.2003;12(1):1-13.

[58] Tsui NB, Kadir RA, Chan KC, Chi C, Mellars G, Tuddenham EG, Leung TY, Lau TK, Chiu RW, Lo YM. Noninvasive prenatal diagnosis of hemophilia by microfluidics digital PCR analysis of maternal plasma DNA. Blood. 2011;117(13):3684-3691. DOI:10.1210/ jc.2014-1118

[59] Volpi EV, Bridger JM. Fish glossary: an overview of the fluorescence in situ hybridization technique. Biotechniques. 2008;45(4):385-386.

[60] Eshaque B, Dixon B. Technology platforms for molecular diagnosis of cystic fibrosis. Biotechnological Adv. 2006;24(1):86-93.

[61] Robinson B, Erle D, Jones D, Shapiro S, Metzger W, Albelda S, Parks W, Boylan A. Recent advances in molecular biological techniques and their relevance to pulmonary research. Thorax. 2000;55(4):329-339.

[62] Singh A, Kumar N. A review on DNA microarray technology. Int J Curr. Res. Rev. 2013;5(22):01-05.

[63] Sato T, Takayanagi A, Nagao K, Tomatsu N, Fukui T, Kawaguchi M, Kudoh J, Amagai M, Yamamoto N, Shimizu N. Simple PCR based DNA microarray system to identify human pathogenic fungi in skin. J. Clinical Microbiol. 2010;48(7):2357-2364.

[64] Zhang N, McCarthy ML, Smart CD. A macroarray system for the detection of fungal and oomycete pathogens of solanaceous crops. Plant Dis. 2008;92(6):953-960

[65] Ahmed OB, Asghar AH, Elhassan MM. Comparison of three DNA extraction methods for polymerase chain reaction (PCR) analysis of bacterial genomic DNA. African J. Microbiol. Res. 2014;8(6):598-602. DOI:10.5897/AJMR2013.6459.

[66] Kuhn J, Binder S. RT–PCR analysis of 5′ to 3′-end-ligated mRNAs identifies the extremities of cox2 transcripts in pea mitochondria. Nucleic Acids Res. 2002;30(2):439-446. Doi:10.1093/ nar/30.2.439.

[67] Das A, Babiuk S, McIntosh MT. Development of a loop-mediated isothermal amplification assay for rapid detection of capripox viruses. J. Clinical Microbiol. 2012;50(5):1613-1620.

[68] Fakruddin MD. Loop mediated isothermal amplification (LAMP)–an alternative to polymerase chain reaction (PCR). Bangladesh Res. Publications J. 2011;5(4): 425-439 .

[69] Liu W, Huang S, Liu N, Dong D, Yang Z, Tang Y, Ma W, He X, Ao D, Xu Y. Establishment of an accurate and fast detection method using molecular beacons in loop-mediated isothermal

amplification assay. Sci. Rep. 2017; 7:40125. doi:10.1038/srep40125 (2017).

[70] Ren CH, Hu CQ, Luo P, Wang QB. Sensitive and rapid identification of Vibrio vulnificus by loop-mediated isothermal amplification. Microbiol. Res. 2009;164(5):514-521.

[71] Villari C, Tomlinson JA, Battisti A, Boonham N, Capretti P, Faccoli M. Use of loop-mediated isothermal amplification for detection of Ophiostoma clavatum, the primary blue stain fungus associated with Ips acuminatus. Appl. Environ. Microbiol. 2013;79(8):2527-2533.

[72] Chang SS, Hsieh WH, Liu TS, Lee SH, Wang CH, Chou HC, Yeo YH, Tseng CP, Lee CC. Multiplex PCR system for rapid detection of pathogens in patients with presumed sepsis–a systemic review and meta-analysis. PloS one. 2013;8(5):e62323. DOI:10.1371/journal.pone.0062323

[73] Lee D, La Mura M, Allnutt TR, Powell W. Detection of genetically modified organisms (GMOs) using isothermal amplification of target DNA sequences. BMC Biotechnol. 2009;9(1):7.

[74] Lizardi PM, Huang X, Zhu Z, Bray-Ward P, Thomas DC, Ward DC. Mutation detection and single-molecule counting using isothermal rolling-circle amplification. Nature Genetics. 1998;19(3):225-232.

[75] Sergentet-Thevenot D, Montet MP, Vernozy-Rozand C. Challenges to developing nucleic acid sequence-based amplification technology for the detection of microbial pathogens in food. Rev. Med. Vet. 2008;159:514-527.

[76] Chen X, Steed A, Harden C, Nicholson P. Characterisation of *Arabidopsis thaliana–Fusarium graminearum* interactions and identification of variation in resistance among ecotypes. Mol. Plant Pathol. 2006;7:391-403.

[77] Ishii H, Holloman DW. Fungicide resistance in plant pathogens. Tokyo: Springer; 2015.

[78] Nakayashiki T, Kurtzman CP, Edskes HK, Wickner RB. Yeast prions [URE3] and [PSI+] are diseases. Proc. Natl. Acad. Sci. 2005;102(30):10575-10580.

[79] Panwar V, McCallum B, Bakkeren G. Endogenous silencing of Puccinia triticina pathogenicity genes through in planta-expressed sequences leads to the suppression of rust diseases on wheat. Plant J. 2012;73:521-532. doi:10.1111/tpj.12047.

[80] de Jonge R, van Esse HP, Maruthachalam K, Bolton MD, Santhanam P, Saber MK, Zhang Z, Usami T, Lievens B, Subbarao KV. Tomato immune receptor Ve1 recognizes effector of multiple fungal pathogens uncovered by genome and RNA sequencing. Proc. Natl. Acad. Sci. 2012;109:5110-5115

[81] Grabherr MG, Haas BJ, Yassour M, Levin JZ, Thompson DA, Amit I, Adiconis X, Fan L, Raychowdhury R, Zeng Q. Full-length transcriptome assembly from RNA-Seq data without a reference genome. National Biotech. 2011;29:644-652.

[82] Mortazavi A, Williams BA, McCue K, Schaeffer L, Wold B. Mapping and quantifying mammalian transcriptomes by RNA-Seq. Nat Methods. 2008;5:621-628.

[83] Metzker ML. Sequencing technologies – the next generation. Nat. Rev. Genet. 2009;11:31-46.

[84] Belete T. Boyraz N. Biotechnological tools for detection, identification and management of plant diseases. African J. Biotech. 2019;18(29):797-807. DOI: 10.5897/AJB2018.16591

[85] Balodi R, Bisht S, Ghatak A, Rao KH. Plant disease diagnosis: technological advancements and challenges. Indian Phytopath. 2017;70(3):275-281.

[86] White TJ, Bruns TD, Lee SB, Taylor JW. Amplification and direct sequencing of fungal ribosomal RNA genes for phylogenetics. In: Innis MA, Gelfand DH, Sninsky JJ, White TJ (eds). PCR protocols: a guide to methods and applications. Academic Press, New York; 1990. 315-322 p. DOI: 10.1016/B978-0-12-372180-8.50042-1.

[87] Werres S, Steffens C. Immunological techniques used with fungal plant pathogens—aspects of antigens, antibodies and assays for diagnosis. Ann. Appl. Biol. 1994;125:615-643.

[88] Hansen K, Lobuglio KF, Pfister DH. Evolutionary relationships of the cup-fungus genus *Peziza* and *Pezizaceae* inferred from multiple nuclear genes: RPB2, beta-tubulin, and LSU rDNA. Mol. Phylogenetics Evo. 2005;36:1-23.

[89] Hyde KD, Jones EBG, Lui J-K, Ariyawansa H, Boehm E, et al. Families of *Dothideomycetes*. Fungal Divers. 2013;63:1-313.

[90] Wang HK, Aptroot A, Crous PW, Hyde KD, Jeewon R. The polyphyletic nature of *Pleosporales*: an example from *Massariosphaeria* based on rDNA and RPB2 gene phylogenies. Mycol. Res. 2007;111:1268-1276

[91] Caporaso JG, Kuczynski J, Stombaugh J et al. QIIME allows analysis of high-throughput community sequencing data. Nat Methods. 2010;7:335-336.

[92] Bates ST, Ahrendt S, Bik HM, Bruns TD, Caporaso JG, Cole J, Dwan M, et al. Meeting report: fungal ITS workshop (October 2012). Stand Genomic Sci. 2013;8:118-123.

[93] Schoch CL, Robbertse B, Robert V, Vu D, Cardinali G, Irinyi L, Meyer W, Nilsson RH, Hughes K, Miller N, Kirk PM. Finding needles in haystacks: linking scientific names, reference specimens and molecular data for Fungi. Database. 2014. DOI:10.1093/database/bau061.

[94] Lanzén A, Jørgensen SL, Huson DH, GorferM, Grindhaug SH, Jonassen I, Øvreås L, Urich T. CREST – classification resources for environmental sequence tags. PLoS One. 2012;7:e49334.

[95] Altschul SF, Madden TL, Schäffer AA, Zhang J, Zhang Z, Miller W, Lipman DJ. Gapped BLAST and PSI-BLAST: a new generation of protein database search programs. Nucleic Acids Res. 1997;25:3389-3402.

[96] Olson A, Aerts A, Asiegbu F, Belbahri L, Bouzid O, Broberg A. et al. Insight into trade-off between wood decay and parasitism from the genome of a fungal forest pathogen. New Phycologist. 2012;194:1001-1013. DOI:10.1111/j.1469-8137. 2012.04128.x.

[97] Notomi T, Okayama H, Masubuchi H, Yonekawa T,Watanabe K, Amino N, Hase T. Loop-mediated isothermal amplification of DNA. Nucleic Acids Res. 2000;28:e63. DOI:10.1093/nar/28.12.e63.

[98] Aglietti C, Luchi N, Pepori AL, Bartolini P, Pecori F, Raio A, Capretti P, Santini A. Real-time loop-mediated isothermal amplification: an early-warning tool for quarantine plant pathogen detection. AMB Express. 2019a;9:50-14. DOI:10.1186/s13568-019-0774-9

[99] Aglietti C, Stehliková D, Paap T, Luchi N, Pecori F, Santini AA. new loop mediated isothermal amplification assay based on assimilating probe for early sequence-specific detection of *Fusarium circinatum* and *F. euwallaceae*. Joint

Meeting of the IUFRO, Working Parties "Shoot, foliage and stem diseases" and "Wilt diseases" (7.02.02 and 7.02.03) on Phyllosphere Diseases. 6-10 May 2019, Figline Valdarno, Florence, Italy, 2019b 49 p.

[100] Harrison C, Tomlinson J, Ostoja-Starzewska S, Boonham N. Evaluation and validation of a loop-mediated isothermal amplification test kit for detection of *Hymenoscyphus fraxineus*. Eur. J.Plant Pathol. 2017;149:253-259. DOI:1007/s10658-017-1179-8

[101] Kav NN, Srivastava S, Yajima W, Sharma N. Application of proteomics to investigate plant-microbe interactions. Current Proteomics. 2007;4(1):28-43.

[102] Klose J. Protein mapping by combined isoelectric focusing and electrophoresis of mouse tissues. Humangenetik. 1975;26(3):231-243.

[103] O'Farrell PH. High resolution two-dimensional electrophoresis of proteins. J. Biol. Chem. 1975;250(10):4007-4021.

[104] Komatsu S, Kamal AH, Hossain Z. Wheat proteomics: proteome modulation and abiotic stress acclimation. Frontiers in plant sci. 2014;5:684.

[105] Barker I. Serological methods in crop protection, in Diagnostics in crop production, BCPC Symposium Proceedings No 65, In: Marshall G, British Crop Protection Council, Farnham, UK; 1996. pp 13-22.

[106] Dewey FM, Thornton CR, Gilligan CA. The use of monoclonal antibodies to detect, quantify and visualise fungi in soils. Adv. Bot. Res. 1997;24:275-308

[107] Dewey FM, Thornton CR. Fungal immunodiagnosis in plant agriculture. 1995.

[108] Halk EL, De Boer SH. Monoclonal antibodies in plant disease research. Ann. Rev. Phytopathol. 1985;23:321-350. DOI:10. 1146/annurev. py.23.090185.001541.

[109] Torrance L. Use of monoclonal antibodies in plant pathology. Europion J. plant Pathol. 1995;101:351-363. DOI:10.1007/BF01874849.

[110] Boonham N, Glover R, Tomlinson J, Mumford R. Exploiting generic platform technologies for the detection and identification of plant pathogens. Eur. J. Plant Pathol. 2008;121:355-363. DOI:10.1007/s10658-008-9284-3.

[111] Tomlinson JA, Dickinson M, Hobden E, Robinson S, Giltrap PM, Boonham N. A five-minute DNA extraction method for expedited detection of *Phytophthora ramorum* following prescreening using *Phytophthora* spp. lateral flow devices. J.Microbiol. Methods. 2010;81:116-120. DOI:10.1016/j.mimet.

[112] Luchi N, Loos R, Santini A. Fast and reliable molecular methods to detect fungal pathogens in woody plants. Appl. Microbiol. Biotechnol. 2020;104:2453-2468. DOI:10.1007/s00253-020-10395-4.

[113] Otrosina WJ, Ghase TE, Gobb FW, Korhonen K. Allozyme differentiation of intersterility groups of *Heterobasidion annosum* isolated from conifers in the Western United States. Phytopathol. 1992;82:540-545. DOI:10.1094/Phyto-82-540.

[114] Raddi S, Santini A, Casini N. Comparison of enzymatic polymorphism in different *Seiridium* isolates. Proc Congr Medit Phytopathology Union, 9th, Kusadasi, Izmir: J. Turkish phytopath. 1994;281-285.

[115] McNeill, J. Barrie FR, Buck WR, Demoulin V, Greuter W, Hawksworth DL, Herendeen PS, Knapp S, Marhold K, Prado J,

Prud'homme Van Reine WF. International Code of Nomenclature for algae, fungi, and plants (Melbourne Code) adopted by the Eighteenth International Botanical Congress Melbourne, Australia, July 2011. [Regnum Vegetabile No. 154.] Königsten: Koeltz Scientific Books; 2012.

[116] Dayarathne MC, Boonmee S, Braun U, Crous PW, Daranagama DA, Dissanayake AJ, Ekanayaka H, Jayawardena R, Jones EBG, Maharachchikumbura SSN, Perera RH, Phillips AJL, Stadler M, Thambugala KM, Wanasinghe DN, Zhao Q, Hyde KD, Jeewon R. Taxonomic utility of old names in current fungal classification and nomenclature: Conflicts, confusion & clarifications. Mycosphere. 2016;7(11):1622-1648. DOI: 10.5943/mycosphere/7/11/2.

[117] Hawksworth DL. A new dawn for the naming of fungi: impacts of decisions made in Melbourne in July 2011 on the future publication and regulation of fungal names. MycoKeys. 2011,7-20. DOI:10.5598/imafungus.2011.02.02.06.

[118] Guarro J, Gené J, Stchigel AM. Developments in fungal taxonomy. Clinical Microbiology Reviews. 1999;12(3):454-500. DOI: 10.1128/CMR.12.3.454.

[119] Wingfield MJ, De Beer ZW, Slippers B, Wingfield BD, Groenewald JZ, Lombard L, Crous PW. One fungus, one name promotes progressive plant pathology. Mol. plant pathol. 2012;13(6):604-613. DOI:10.1111/j.1364-3703.2011.00768.x.

[120] Lombard L, Crous PW, Wingfield BD, Wingfield MJ. Multigene phylogeny and mating tests reveal three cryptic species related to *Calonectria pauciramosa*. Stud. Mycol. 2010a;66:15-30. DOI:10.3114/sim.2010.66.02.

[121] Gams W, Jaklitsch W. A critical response to the 'Amsterdam Declaration'. Mycotaxon. 2011;116:501-512. DOI:10.5598/imafungus.2013.04.01.06.

[122] Hawksworth DL, McNeill J, de Beer W, Wingfield MJ. Names of fungal species with the same epithet applied to different morphs: how to treat them. IMA fungus. 2013;4(1):53-56. DOI:10.5598/imafungus.2013.04.01.06.

[123] de Beer ZW, Duong TA, Barnes IBD, Wingfield BD, Wingfield MJ. Redefining *Ceratocystis* and allied genera. Studies in Mycology. 2014;79:187-219. DOI:10.1016/j.simyco.2014.10.001.

[124] Jankowiak R, Strzałka B, Bilański P, Linnakoski R, Aas T, Solheim H, Groszek M, de Beer ZW. Two new *Leptographium* spp. reveal an emerging complex of hardwood-infecting species in the Ophiostomatales. Antonie Van Leeuwenhoek. 2017;110:1537-1553. DOI:10.1007/s10482-017-0905-8.

[125] Taylor JW. One Fungus = One Name: DNA and fungal nomenclature twenty years after PCR. IMA Fungus. 2011;2:113-120. DOI:10.5598/imafungus.2011.02.02.01.

[126] Carnegie AJ, Cooper K. Emergency response to the incursion of an exotic myrtaceous rust in Australia. Australasian Plant Pathol. 2011;40:346-359. DOI:10.1007/s13313-011-0066-6.

[127] Coutinho TA, Wingfield MJ, Alfenas AC, Crous PW. *Eucalyptus* rust: a disease with the potential for serious international implications. Plant Dis. 1998;82:819-825. DOI:10.1094/PDIS.1998.82.7.819.

[128] Glen M, Alfenas AC, Zauza EA, Wingfield MJ, Mohammed C. *Puccinia psidii*: a threat to the Australian environment and economy–a review. Australasian Plant Pathol. 2007;36(1):1-6. DOI:10.1071/AP06088.

[129] Gams W. A new nomenclature for fungi. Mycologia Iranica. 2014;1(1):1-5. DOI:10.22043/MI.2014.2959.

[130] Rossman A, Seifert K. Phylogenetic revision of taxonomic concepts in the Hypocreales and other Ascomycota-a tribute to Gary J. Samuels. Stud. Mycol. 2011;68.

[131] Gräfenhan T, Schroers HJ, Nirenberg HI, Seifert KA. An overview of the taxonomy, phylogeny, and typification of nectriaceous fungi in *Cosmospora*, *Acremonium*, *Fusarium*, *Stilbella*, and *Volutella*. Studies in Mycology. 2011;68:79-113. DOI:10.3114/sim.2011.68.04.

[132] Summerbell RC, Gueidan C, Schroers HJ, De Hoog GS, Starink M, Rosete YA, Guarro J, Scott JA. *Acremonium* phylogenetic overview and revision of *Gliomastix*, *Sarocladium*, and *Trichothecium*. Stud. Mycol. 2011;68:139-162. DOI:10.3114/SIM.2011.68.06.

[133] Chaverri P, Salgado C, Hirooka Y, Rossman AY, Samuels GJ. Delimitation of *Neonectria* and *Cylindrocarpon* (Nectriaceae, Hypocreales, Ascomycota) and related genera with Cylindrocarpon-like anamorphs. Studies in Mycology. 2011;68:57-78. Doi:10.3114/sim.2011.68.03.

[134] Hirooka Y, Rossman AY, Chaverri P. A morphological and phylogenetic revision of the *Nectria cinnabarina* species complex. Studies in Mycol. 2011;68:35-56. Doi:10.3114/sim.2011.68.02.

[135] Mejía LC, Castlebury LA, Rossman AY, Sogonov MV, White Jr JF. A systematic account of the genus *Plagiostoma* (Gnomoniaceae, Diaporthales) based on morphology, host-associations, and a four-gene phylogeny. Stud. Mycol. 2011;68:211-235. Doi:10.3114/sim.2011.68.10.

[136] Sultan A, Johnston PR, Park D, Robertson AW. Two new pathogenic ascomycetes in *Guignardia* and *Rosenscheldiella* on New Zealand's pygmy mistletoes (*Korthalsella*: Viscaceae). Stud. Mycol. 2011;68:237-247. Doi:10.3114/sim.2011.68.11.

[137] Põldmaa K. Tropical species of *Cladobotryum* and *Hypomyces* producing red pigments. Stud. Mycol. 2011;68:1-34. DOI:10.3114/sim.2011.68.01.

[138] Réblová M, Seifert KA. Discovery of the teleomorph of the hyphomycete, *Sterigmatobotrys macrocarpa*, and epitypification of the genus to holomorphic status. Stud. Mycol. 2011;68:193-202. DOI:10.3114/sim.2011.68.08.

[139] Crous PW, Groenewald JZ, Shivas RG, Edwards J, Seifert KA, Alfenas AC, Alfenas RF, Burgess TI, Carnegie AJ, Hardy GS, Hiscock N. Fungal Planet description sheets: 69-91. Persoonia, 2011; 26:108-156. DOI: 10.3767/003158511x581723 PMID: 22025808 PMCID: PMC3160798

[140] Lombard L, Crous PW, Wingfield BD, Wingfield MJ. Phylogeny and systematics of the genus *Calonectria*. Stud. Mycol. 2010b;66:31-69. DOI:10.3114/sim.2010.66.03.

[141] Lombard L, Crous PW, Wingfield BD, Wingfield MJ. species concepts in *Calonectria* (*Cylindrocladium*). Stud. Mycol. 2010c;66:1-3. DOI:10.3114/sim.2010.66.01.

[142] Lombard L, Rodas CA, Crous PW, Wingfield BD, Wingfield MJ. *Calonectria* (*Cylindrocladium*) species associated with dying Pinus cuttings. Persoonia. 2009;23:41. DOI:10.3767/003158509X471052.

[143] Aylward J, Steenkamp ET, Dreyer LL, Roets F, Wingfield BD, Wingfield MJ. A plant pathology perspective of fungal genome

sequencing. IMA fungus. 2017;8(1):1-5 DOI:10.5598/imafungus.2017.08.01.01.

[144] Braun U, Cook RTA. Taxonomic Manual of the Erysiphales (Powdery Mildews). CBS Biodiversity Series. 2012; 11:1-707.

[145] Hawksworth DL. Global species numbers of fungi: Are tropical studies and molecular approaches contributing to a more robust estimate? Biodivers. Conserv. 2012;21:2425-2433.

[146] Peterson SW. Phylogenetic relationships in *Aspergillus* based on rDNA sequence analysis. In: Integration of Modern Taxonomic Methods for *Penicillium* and *Aspergillus* Classification, Samson RA, Pitt JI, eds. Amsterdam: Harwood Academic Publishers; 2000. 323-356 p.

[147] Taylor JW, Jacobson DJ, Kroken S, Kasuga T, Geiser DM, Hibbett DS, Fisher MC. Phylogenetic species recognition and species concepts in fungi. Fungal Genet. Biology. 2000;31(1):21-32. DOI:10.1006/fgbi.2000.1228.

[148] Taylor JW, Göker M, Pitt JI. Choosing one name for pleomorphic fungi: the example of *Aspergillus* versus *Eurotium*, *Neosartorya* and *Emericella*. Taxon. 2016;65(3):593-601. DOI:10.12705/653.10.

[149] Kirk PM, Stalpers JA, Braun U, Crous PW, Hansen K, Hawksworth DL, Hyde KD, Lücking R, Lumbsch TH, Rossman AY, Seifert KA. A without-prejudice list of generic names of fungi for protection under the International Code of Nomenclature for algae, fungi, and plants. IMA fungus. 2013;4(2):381-443. DOI:10.5598/imafungus.2013.04.02.17.

[150] Rifai MA. A Revision of the Genus *Trichoderma*. Mycological Papers. 1969;116:1-56.

[151] Druzhinina IS, Kubicek CP, Komoń-Zelazowska M, Mulaw TB, Bissett J. The *Trichoderma harzianum* demon: complex speciation history resulting in coexistence of hypothetical biological species, recent agamo species and numerous relict lineages. BMC Evolutionary Biology. 2010;10(1):1-4. DOI:10.1186/1471-2148-10-94.

[152] Chaverri P, Branco-Rocha F, Jaklitsch W, Gazis R, Degenkolb T, Samuels GJ. Systematics of the *Trichoderma harzianum* species complex and the re-identification of commercial biocontrol strains. Mycologia. 2015;107(3):558-590. DOI:10.3852/14-147.

[153] Li QR, Tan P, Jiang YL, Hyde KD, Mckenzie EH, Bahkali AH, Kang JC, Wang Y. A novel *Trichoderma* species isolated from soil in Guizhou, *T. guizhouense*. Mycol. Prog. 2013;12(2): 167-172. DOI:10.1007/s11557-012-0821-2.

[154] Woudenberg JH, Groenewald JZ, Binder M, Crous PW. *Alternaria* redefined. Stud. Mycol. 2013;75:171-212. DOI:10.3114/sim0015.

[155] Hyde KD, Nilsson RH, Alias SA, Ariyawansa HA, Blair JE, Cai L, de Cock AW, Dissanayake AJ, Glockling SL, Goonasekara ID, Gorczak M. One stop shop: backbones trees for important phytopathogenic genera: I (2014). Fungal Divers. 2014;67(1):21-125. DOI:10.1007/s13225-014-0298-1.

[156] Jayawardena RS, Hyde KD, Jeewon R, Ghobad-Nejhad M, Wanasinghe DN, Liu N, Phillips AJ, Oliveira-Filho JR, da Silva GA, Gibertoni TB, Abeywikrama P. One stop shop II: taxonomic update with molecular phylogeny for important phytopathogenic genera: 26-50 (2019a). Fungal Divers. 2019a;94(1):41-129. DOI:10.1007/s13225-019-00418-5.

[157] Jayawardena RS, Hyde KD, McKenzie EH, Jeewon R, Phillips AJ, Perera RH, de Silva NI, Maharachchikumburua SS, Samarakoon MC, Ekanayake AH, Tennakoon DS. One stop shop III: taxonomic update with molecular phylogeny for important phytopathogenic genera: 51-75 (2019). Fungal Divers. 2019b;98(1):77-160. DOI: 10.1007/s13225-019-00433-6(0123456789.

[158] Jayawardena RS, Hyde KD, McKenzie EHC. et al. One stop shop IV: taxonomic update with molecular phylogeny for important phytopathogenic genera: Fungal Diversity. 2020;99:76-100. DOI:10.1007/s13225-020-00460-8.

[159] Marin-Felix Y, Groenewald JZ, Cai L, Chen Q, Marincowitz S, Barnes I, Bensch K, Braun U, Camporesi E, Damm U, de Beer ZW, Dissanayake A, Edwards J, Giraldo A, Hernandez-Restrepo M, Hyde KD, Jayawardena RS, Lombard L, Luangsa-ard J, McTaggart AR, Rossman AY, Sandoval-Denis M, Shen M, Shivas RG, Tan YP, van der Linde EJ, Wingfield MJ, Wood AR, Zhang JQ, Zhang Y, Crous PW. Genera of phytopathogenic fungi: GOPHY 1. Stud. Mycol. 2017;86:99-216. Doi:10.1016/j.simyco.2017.04.002.

[160] Marin-Felix Y, Hernández-Restrepo M, Iturrieta-González I, García D, Gené J, Groenewald JZ, Cai L, Chen Q, Quaedvlieg W, Schumacher RK, Taylor PWJ, Ambers, C, Bonthond G, Edwards J, Krueger-Hadfield SA, Luangsa-ard JJ, Morton L, Moslemi A, Sandoval-Denis M, Tan YP, Thangavel R, Vaghefi N, Cheewangkoon R, Crous PW. Genera of phytopathogenic fungi: GOPHY 3. Stud. Mycol. 2019b;94:1-124. Doi:10.1016/j.simyco.2019.05.001.

[161] Marin-Felix Y, Hernández-Restrepo M, Wingfield MJ, Akulov A, Carnegie AJ, Cheewangkoon R, Gramaje D, Groenewald JZ, Guarnaccia V, Halleen F, Lombard L, Luangsa-ard J, Marincowitz S, Moslemi A, Mostert L, Quaedvlieg W, Schumacher RK, Spies CFJ, Thangavel R, Taylor PWJ, Wilson AM, Wingfield BD, Wood AR, Crous PW. Genera of phytopathogenic fungi: GOPHY 2. Stud. Mycol. 2019a;92:47-133. DOI:10.1016/j.simyco.2018.04.002

[162] Vu D, Groenewald M, De Vries M, Gehrmann T, Stielow B, Eberhardt U, Al-Hatmi A, Groenewald JZ, Cardinali G, Houbraken J, Boekhout T. Large-scale generation and analysis of filamentous fungal DNA barcodes boosts coverage for kingdom fungi and reveals thresholds for fungal species and higher taxon delimitation. Stud. Mycol. 2019;92:135-154. DOI:10.1016/j.simyco.2018.05.001.

[163] Seifert KA. Progress towards DNA barcoding of fungi. Mol. Ecol. Resources. 2009;9:83-89. DOI:10.1111/j.1755-0998.2009.02635.x

[164] Goldstein PZ, DeSalle R. Integrating DNA barcode data and taxonomic practice: determination, discovery, and description. Bioessays. 2011;33(2):135-147. DOI:10.1002/bies.201000036.

[165] Hawksworth DL, Hibbett DS, Kirk PM, Lücking R. (308-310) Proposals to permit DNA sequence data to serve as types of names of fungi. Taxon. 2016;65(4):899-900. DOI:10.12705/654.31.

[166] Hawksworth DL, Hibbett DS, Kirk PM, Lücking R. (F-005-006) Proposals to permit DNA sequence data to be used as types of names of fungi. IMA Fungus. 2018;9:v–vi.

[167] Zamora JC, Svensson M, Kirschner R, Olariaga I, Ryman S, Parra LA, Geml J, Rosling A, Adamčík S, Ahti T, Aime MC.

Considerations and consequences of allowing DNA sequence data as types of fungal taxa. IMA fungus. 2018;9(1):167-175. DOI:10.5598/imafungus.2018.09.01.10.

[168] Callahan BJ, McMurdie PJ, Holmes SP. Exact sequence variants should replace operational taxonomic units in marker-gene data analysis. The ISME J. 2017;11(12):2639-2643. DOI:10.1038/ismej.2017.119.

Chapter 3

UAV Remote Sensing: An Innovative Tool for Detection and Management of Rice Diseases

Xin-Gen Zhou, Dongyan Zhang and Fenfang Lin

Abstract

Unmanned aerial vehicle (UAV) remote sensing is a new alternative to traditional diagnosis and detection of rice diseases by visual symptoms, providing quick, accurate and large coverage disease detection. UAV remote sensing offers an unprecedented spectral, spatial, and temporal resolution that can distinguish diseased plant tissue from healthy tissue based on the characteristics of disease symptoms. Research has been conducted on using RGB sensor, multispectral sensor, and hyperspectral sensor for successful detection and quantification of sheath blight (*Rhizoctonia solani*), using multispectral sensor to accurately detect narrow brown leaf spot (*Cercospora janseana*), and using infrared thermal sensor for detecting the occurrence of rice blast (*Magnaporthe oryzae*). UAV can also be used for aerial application, and UAV spraying has become a new means for control of rice sheath blight and other crop diseases in many countries, especially China and Japan. UAV spraying can operate at low altitudes and various speeds, making it suitable for situations where arial and ground applications are unavailable or infeasible and where precision applications are needed. Along with advances in digitalization and artificial intelligence for precision application across fertilizer, pest and crop management needs, this UAV technology will become a core tool in a farmer's precision equipment mix in the future.

Keywords: rice, UAV, drone, sensor, remote sensing, vegetation index, PCR, sheath blight, blast, narrow brown leaf spot, *Rhizoctonia solani*, *Magnaporthe oryzae*, *Cercospora janseana*, symptoms, fungicides, threshold for economical fungicide application

1. Introduction

Rice (*Oryza sativa* L.) is one of the three major food crops (rice, wheat, and maize) with worldwide production of 502 million tons, feeding more than half of the world's human population. China and the US rank 1st and 17th in worldwide rice production, producing 149 million and 7 million tons of rice in 2020, respectively [1]. Rice is one of major field crops in the US and is planted on 1.3 million hectares in Arkansas, California, Mississippi, Missouri, and Texas.

However, the occurrence of diseases poses a major threat to rice production. Numerous fungal, bacterial, viral, and nematode diseases occur in rice-growing regions, causing significant yield and quality losses annually [2]. Rice blast caused by *Magnaporthe oryzae* (formerly *M. grisea*) is the most important disease worldwide followed by sheath blight caused by *Rhizoctonia solani* AG1-1A. In the US rice, sheath blight is the number 1 disease, causing more economic loss than rice blast [3–5]. Sheath blight infects leaf sheaths, leaf blades, and even panicles, causing up to 44% yield loss and a significant reduction in milling quality [6]. Narrow brown leaf spot (NBLS), caused by *Cercospora janseana*, is another important disease with worldwide distributions [7]. NBLS, considered a minor disease historically, has become one of the most important rice diseases in the southern US, especially Texas and Louisiana with the humid, warm Gulf Coast climate. NBLS is more severe at late plantings and in the ratoon (second) crop. Ratooning is a practice following main crop harvest to maximize production returns in Texas and Louisiana where cropping season is long enough for 2nd crop harvest. NBLS attacks leaf, sheath, internode, panicle branch and glume tissues, causing characteristic linear brown lesions. NBLS can cause grain yield loss of up to 40% [7].

Rapid and accurate identification and detection of rice disease are the first essential step for effective management of these diseases. Diagnosis of diseases by their visual characteristic symptoms is the most common practice at present. However, this disease detection process is time consuming and labor-some. Accuracy of disease identification and detection is highly dependent on the knowledge and experience of the inspector. For example, to detect and monitor sheath blight, visual inspection should start at the panicle differentiation growth stage of rice by walking across the field in a zigzag pattern many times [8]. Many stops are needed to scout for the presence of the disease based on its symptoms in the lower portion of canopy. This process repeats weekly at the early stages and more frequently (biweekly) at the late stages or under conditions most favorable for sheath blight to develop until heading.

Unmanned aerial vehicle (UAV) remote sensing provides a new way to detect and monitor disease development. UAV remote sensing offers a quick, accurate, large area coverage, and low-cost tool for disease assessment. Remote sensing is the science and art of acquiring information about material objects from measurements made at a distance without coming in physical contact with the objectives [9]. Remote sensors can sense the changes in spectral reflectance that results from the changes of external biophysical and internal biochemical characteristics of plant tissue [10]. Spectral properties of vegetation are determined by plant tissue features, including pigment and moisture content of tissue, leaf area index, ratio of live and senesced tissue biomass, and spatial arrangement of cells and structures [11]. Changes in the spectral properties of vegetation occur in the three distinguished spectral domains of vegetation reflectance, invisible (VIS: 0.4 to 0.7 μm), near-infrared (NIR: 0.7 to 1.3 μm), and mid-infrared (mid-IR: 1.3 to 2.5 μm) [11]. Infected or diseased plants change their spectral properties of vegetation. A reduction in chlorophyll production in infected tissue results in less absorption of blue and red band visible light. These changes are reflected in all the three blue, green, and red bands. So, yellow or brown color is present in infected tissue image. In infected plants, NIR bands are not absorbed by mesophyll cells but by stressed and dead cells, resulting in the presence of dark tones in acquired image. Therefore, remote sensing can detect these changes in spectral reflection pattern in infected or diseased plant tissue. This is the basis for the application of remote sensing on plant disease detection and quantification. Various sensors, including digital RGB (Red, Green and Blue) sensor, multispectral sensor, hyperspectral sensor, fluorescence imaging, and infrared thermal sensors, have been widely utilized to characterize plant disease symptoms, detect different diseases, and even quantify severity of many plant diseases in the laboratory and field [12–14].

UAV can also be used as an aerial fungicide sprayer for disease control. UAV spraying can operate at low altitudes and various speeds, and apply with low volumes, making it suitable for situations where arial and ground applications are unavailable or infeasible and where precision applications are needed for more economically and environmentally effective control of diseases. With the development of UAV technology, the use of UAV for aerial fungicide application has become a new means for control of diseases in rice and other crops in recent years [15]. Considerable acreage of rice crops has been treated with UAV spraying in many countries, especially China and Japan [15–18]. In 2020, total treated areas of crops were over 450 million hectares in China [16]. In Japan, approximately 40% of rice acreage is treated with UAV spraying [17, 18].

This article reviews the recent advances in the research and use of UAV remote sensing for the detection and management of crop diseases, with a focus on sheath blight and NBLS, two important fungal diseases in rice. This review article covers disease symptoms, traditional disease detection methods, remote sensing for disease detection, and UAV used as a tool for disease management. Conclusion and prospects are also included at the end of this review article.

2. Disease symptoms

2.1 Sheath blight

Sheath blight is soilborne disease and the fungus can survive as sclerotia in soil for up to 2 years [19]. The disease starts with the contact of sclerotia with leaf sheaths at or just above the water line after the sclerotia float out of the soil with irrigation water. The sclerotia germinate and infect the leaf sheaths at the stages of later tillering to early reproduction. Initial symptoms develop on the leaf sheaths and are circular, oval or ellipsoid, water-soaked spots in greenish-gray color (**Figure 1A**). The lesions enlarge and coalesce forming bigger lesions with irregular outlines and grayish-white centers surrounded by dark brown borders (**Figure 1A** and **B**). As lesions coalesce on the sheaths, entire leaves eventually die. Lesions on the leaf blades are more irregular with dark green, brown or yellow-orange margins. The lesions can develop extensively and coalesce on partial or whole leaf blades, producing a rattlesnake skin pattern (**Figure 1C**). Sheath blight spreads in the field vertically and horizontally. The disease moves up the plants (**Figure 2**) and may infect the flag leaves and panicles (**Figures 1D** and **3A**) under severe conditions. The fungus spreads in the field by growing its runner hyphae from tiller to tiller, from leaf to leaf, and from plant to plant, resulting in a circular pattern of damage (**Figure 3A**). The fungus can spread into the culms from early sheath infections and weaken the infected culms, resulting in the lodging and collapse of tillers (**Figure 3B**). Diseased plants reduce grain filling, especially in the lower portion of the panicles. Losses in yield tend to be more severe with increased lodging [8].

Sclerotia, the survival structures of the fungus, form on the surfaces of some sheaths and leaf blades. The sclerotia are white (**Figure 1E**) when first formed, and then turn brown or dark brown (**Figure 1F**). The sclerotia fall off the plants and serve as primary inoculum the following season. Mycelia in infect plant debris can also service as primary inoculum. Sheath bight is considered a monocyclic disease since the pathogen infection cycle occurs only once during a cropping season. The fungus does not produce any asexual or sexual spores for repeated infections under the field conditions [20].

Figure 1.
Sheath blight lesions on the sheaths (A and B), leaf (C) and flag leaf (D), and white (E) and dark (F) sclerotia on the sheaths.

Figure 2.
Vertical development of sheath blight in pathogen-inoculated field plot (A) in comparison with healthy plants in non-inoculated field plot (B) at Eagle Lake, Texas, USA.

2.2 Narrow brown leaf spot (NBLS)

The NBLS fungus is seedborne and survives in infected seed and rice plant residue year after year, serving as primary inoculum. The fungus produces conidia, the structures for infection. Infection starts when the conidia germinate and penetrate host plant tissue through the stomata and grow intercellularly in the tissue [7]. The fungus attacks the leaves (**Figure 4A**), sheaths (**Figure 4B**), internodes, panicle branches and glumes (**Figure 4D**). On leaf blades, it causes short, linear, narrow, brown lesions parallel to the leaf veins (**Figure 4A**). Infection of the leaf sheaths results in a large, brown blotch or "net blotch" caused by the browning of the leaf

Figure 3.
Infected flag leaves and panicles in the circular infection area of field research plot (A), and lodging caused by sheath blight in a commercial rice field at Beaumont (B), Texas, USA.

Figure 4.
Narrow brown leaf spot (NBLS) lesions on the leaves (A) and sheath (B); NBLS net blotch symptoms on the sheath (C); NBLS lesions on panicle branches and glumes (D); and NBLS "neck blast" symptoms at the base of the panicle (E).

veins (**Figure 4C**). The fungus also can cause a "neck blight," where the internodal area above and below the node at the base of the panicle becomes light brown to tan (**Figure 4E**). The affected tissue area dies and the kernels in the lower portion of the panicle fail to fill (**Figure 4D**). As plants approach maturity, leaf spotting can become severe on susceptible cultivars and result in severe leaf blighting and premature death (**Figure 5**). The disease can cause premature ripening, yield reduction, and reduced milling quality. Low nitrogen levels increase the severity of the disease. The disease tends to be more severe at late plantings and in ratoon (second) crop. Ratooning is a common practice following main crop harvest in Texas and Louisiana to maximize the returns of rice production [21].

NBLS is a typical polycyclic foliar disease and infection occurs multiple times within a cropping season. The development of symptoms may take 7 days under conducive conditions (Zhou, unpublished data) to 30 days after infection [7].

Figure 5.
Severe narrow brown leaf spot (NBLS) and premature death in the field at Beaumont, Texas, USA.

The disease is airborne and spreads long distances by wind-borne spores, resulting in uniform distribution of the disease in the field [22].

3. Traditional disease detection

3.1 Sheath blight

Rapid and accurate identification and detection of sheath blight are critical for rice growers to employ a right measure such as fungicide application for control of this disease. Diagnosis of sheath blight by visual symptoms is the common practice to detect the presence of the disease. However, diagnosis based on symptoms is difficult since sheath blight may be confused with other sheath diseases with similar symptoms, such as sheath spot (also known bordered sheath spot) caused by *R. oryzae* and aggregate sheath spot caused by *R. oryzae-sativae*, especially at the early stages of disease development. Cooccurrence of these sheath blight-like diseases can be found in many rice production regions of Africa, Asia, North America, and South America [23, 24]. In the US, sheath blight, sheath spot, and aggregate sheath spot are commonly present in the field although sheath spot and aggregate sheath spot usually do not cause measurable yield loss in the southern US [8, 25, 26] whereas sheath blight and sheath spot do not cause yield loss in California [27]. Currently, sheath blight and aggregate sheath spot are the two key rice diseases in the southern states and California, respectively. To improve the accuracy of disease diagnosis, research has been conducted using polymerase chain reaction (PCR) technology for the detection of sheath blight. Matsumoto et al. [23] and Johanson et al. [28] developed PCR-based methods to distinguish these three pathogens. Sayler and Yang [29] developed a real-time PCR assay that can be used to detect and quantify the sheath blight pathogen in infected tissue at the early stages of infection. These molecular methods provide a new tool for the accurate detection of sheath blight in rice. However, this molecular approach has not been adopted for use in commercial rice production in Texas and other United States because of relatively high costs and inaccessibility of this technology to rice farmers. At present, the U.S. farmers still use the traditional diagnosis of sheath blight based on the characteristics of symptoms distinguishable from other diseases as described above.

Scouting for sheath blight and determining the need to trigger fungicide application are important for profitable production of rice. As other diseases, it is difficult to precisely estimate the potential levels of sheath blight in a field to make an assessment on the economic feasibility of applying a fungicide. However, given the high costs of fungicide applications and farmers' need to reduce production costs, the proper disease scouting and assessment is highly recommended. Damage caused by the disease depends on several factors that include cultivar susceptibility, disease pressure, weather conditions, plant density, and nitrogen fertilizers.

Sheath blight develops quickly under favorable environmental conditions. The following field scouting procedure is recommended for the rice farmers in Texas [8]. Detecting and monitoring the development of sheath blight should start scouting for the presence of sheath blight at the panicle differentiation growth stage of rice by walking across the field in a zigzag pattern (**Figure 6**). Farmer should periodically inspect rice plants above the water line for the early symptoms. If there is no sheath blight observed, the farmer should wait a week and monitor again; if some sheath blight is found, a more detailed monitoring procedure should be followed to accurately determine the severity of sheath blight. A large field should be divided into 45 to 50 acres (18 to 20 hectares) sections and inspection made in each section separately (**Figure 6**) to monitor more precisely. The farmer should walk the field sections in a "U" pattern and randomly stop to check for the presence of sheath blight. The stopping point is considered positive for sheath blight even if only one small sheath blight lesion is found on a single plant; the stopping point is considered negative if no sheath blight symptoms is observed. The total number of stops should be at least equivalent to the number of areas scouted (i.e. 50 acres (20 hectares) = 50 or more stops). To the end of the scouting, the percentage of positive sheath blight stopping points can be calculated by dividing the number of positive stops where sheath blight was found by the number of stops and multiplying by 100. Alternatively, the percentage of tillers infected can be calculated by dividing the number of tillers infected by the total number of tillers inspected and multiplying by 100. For the tiller infection assessment, the number of tillers with at least one sheath blight lesion and the total number of tillers inspected should be recorded in each stop.

The thresholds for economical fungicide application are based on the amount of sheath blight present and the susceptibility of the cultivar planted. With very susceptible and susceptible cultivars, 35% positive stops indicate that a fungicide is necessary; moderately susceptible cultivars require 50% positive stops to justify a fungicide application. Alternatively, with very susceptible and susceptible cultivars, 5 to 10% tillers infected indicate that spraying a fungicide is warranted; moderately susceptible cultivars require 10 to 15% tillers infected to justify a fungicide application.

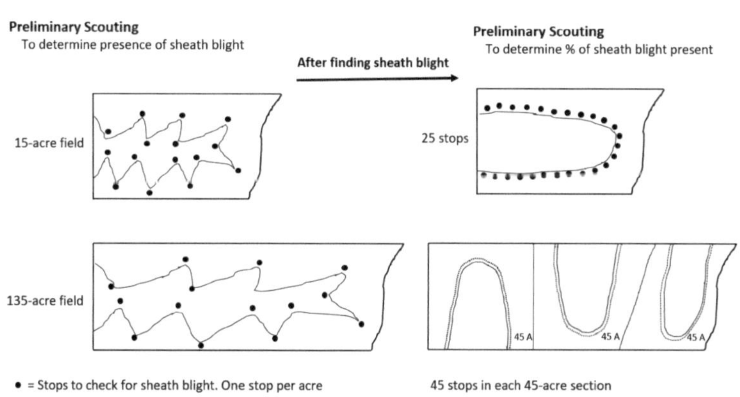

Figure 6.
Scouting procedure for sheath blight detection in the 15-, 45-, and 135-acre (6-, 18-, and 55-hectare) field (source: [8]). A = acre.

This scouting procedure should be repeated until the heading growth stage. However, if most conducive conditions are present and persistent at the growth stages after panicle differentiation, sheath blight should be scouted at the intervals of two times a week.

3.2 Narrow brown leaf spot (NBLS)

NBLS has its own characteristic symptoms that can be distinguished from other diseases in rice. This is especially true when the disease is at its late development stages. However, diagnosis of NBLS by visual symptoms may be confused with other diseases, including brown spot (caused by *Cochliobolus miyabeanus*), white leaf streak (*Mycovellosiella oryzae*), and leaf blast (*Magnaporthe oryzae*). This difficulty becomes more profound when these four diseases are at their initial stages with similar spot symptoms. Like NBLS, brown spot and rice blast, especially brown spot, are commonly present in rice. All rice cultivars, including hybrids, are susceptible to brown spot. Fortunately, white leaf streak is of not much concern for disease diagnosis since it has been reported only in Texas and Louisiana [30, 31] and in several other countries [32]. White leaf streak is a minor disease and is not commonly present in rice.

Scouting for NBLS and monitoring its development are relatively simple compared to sheath blight. NBLS spreads by air-borne spores and its distribution in the field is uniform, which contrasts with sheath blight that is soilborne and spreads in the field in an aggregated pattern. NBLS symptoms first appear on old leaves and then develop on the upper leaves. Rice plants are susceptible to NBLS at all growth stages but become more susceptible from panicle emergence to maturity. Due to relatively slow development of the disease, weekly scouting for NBLS is recommended and the scouting procedure should start at the boot stage until heading.

The thresholds for economical fungicide application have not been established for NBLS yet. However, determining the need to trigger a fungicide application is based on the susceptibility of the cultivar planted, its growth stage, and weather conditions. Significant differences in susceptibility among rice cultivars are present. Some cultivars, especially hybrids, with acceptable levels of resistance do not need a fungicide treatment. A fungicide application may be warranted under the most conducive environments, including combinations of very susceptible cultivars, early growth stages infected, favorable weather conditions and a consideration of ratoon cropping.

4. Remote sensing for disease detection

Remote sensing is an innovation to plant disease detection and monitoring since it provides rapid, accurate and objective observations and can be available real time and all the time. With remote sensing technology, we can rapidly and accurately observe and assess crop growth and disease development at large field scales and make it possible conduct multiple surveys and assessments within a short period of time. This is especially useful when surveying field crops such as rice to cover large field areas. The average size of field crops per farm in the US is relatively large and has continued increase. For example, average rice hectares per U. S. rice farm have been increased from 160 hectares in 2000 to 243 hectares in 2013 [33]. Using traditional field observations and ground surveys for crop diseases, such as rice sheath blight detection method described above, is a challenge to farmers since such visual inspection methods are time-consuming and labor-some. Such manual inspection is also subjective and random, and its accuracy is dependent on the knowledge and experience of the inspector.

4.1 Remote sensors

Various sensors, including digit (RGB) camera, multispectral camera, hyperspectral camera, infrared thermal imager, and fluorescence imaging, have been used in remote sensing for plant disease detection and monitoring [10, 12, 34]. RGB camera is one of the most used sensors because of its light weight, low cost, ease of operation, simple data processing, and low work environment requirements [34]. RBG camera can acquire grayscale or color images, which enables to detect diseased plant tissues with modifications in color, texture, and other spectral information. Hoverer, due to the limitation of fewer visible light bands, RGB camera might provide insufficient spectral information to accurately characterize symptoms and identify diseases. Vegetation features can be identified by extracting color indices from high-resolution images since each pixel value of image can be calculated from the reflectance or radiance of specific bands [34]. RGB camera has been used for the identification and detection of cotton bacterial angular (*Xanthomonas campestris*) and Ascochyta blight (*Ascochyta gossypii*) [35], grapefruit citrus canker (*X. axonopodis*) [36], and sugar beet Cercospora leaf spot (*C. beticola*) and rust (*Uromyces betae*) [37]. In rice, Zhang et al. [38] has successfully used RBG camera to detect and quantify sheath blight. Kurniawati et al. [39] used RGB images to achieve 95% of accuracy to diagnose rice blast, brown spot, and NBLS. Lu et al. [40] identified 10 rice diseases from RGB images using deep conventional neural networks (CNN) with an accuracy of 95%. These 10 rice diseases were rice blast, false smut (*Ustilaginoidea virens*), brown spot, bakanae (*Gibberella fujikuroi*), sheath blight, sheath rot, bacterial leaf blight (*X. oryzae pv. oryzae*), bacterial sheath rot (*Pseudomonas fuscovaginae*), seeding blight, and bacterial wilt.

Multispectral camera is the second most used sensor for plant disease detection. Multispectral sensors can sense and record the radiations from the visible and invisible portions of the electromagnetic spectrum. To the users, multispectral sensors are relatively inexpensive and have the advantages of fast frame imaging and high work efficiency. However, multispectral sensors have their limitations since they have low number of bands, discontinuous spectrum, and low spectral resolution [34]. Multispectral sensing has been used for the classification and detection of more than 16 fungal and bacterial diseases in over 11 field crops [12]. In rice, Zhang et al. [38] reported that multispectral sensor performed better in the detection of sheath blight in field plots compared to RGB sensor. Cai et al. [41] successfully used a multiple spectral sensor to detect and quantify NBLS in the field. Shi et al. [42] used PlanetScope multispectral imaging to classify and detect rice blast, dwarf virus, and glume blight (*Phyllosticta glumarum*) at a large field scale with an accuracy of 76%. Kobayashi et al. [43] evaluated the potential use of multispectral sensor for airborne detection of rice blast.

Hyperspectral sensing is another common method to diagnose and detect plant diseases. Hyperspectral sensors can sense and record a large number of very narrow bands and continuous spectra. They can provide more spectral band information and higher spectral resolution than multispectral sensors. Therefore, hyperspectral imagers have more capacity to capture spectral characteristics of symptoms and crops and to distinguish the differences in spectral traits between different crops [34]. The hyperspectral sensing approach has been used for the detection of more than 12 fungal, bacterial, and nematode diseases in 15 field crops [12]. In rice, sheath blight, rice blast and bacterial leaf blight can be identified with an accuracy of more than 93% using hyperspectral imaging data through machine leaning methods [44]. Most recently, Lin et al. [45] analyzed and compared the spectral responses to rice leaf and sheath tissue infected with sheath blight with healthy tissue and found that the hyperspectral sensing approach performed very well on the

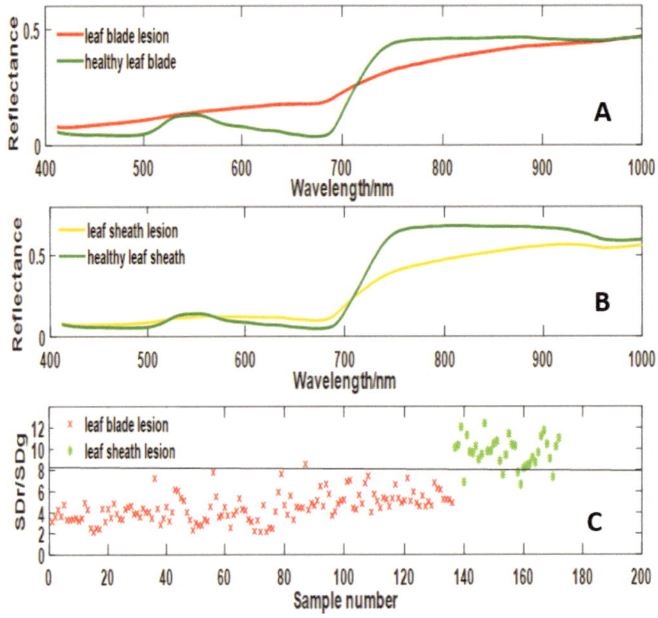

Figure 7.
Comparison of spectral curves of sheath blight-infected rice leaf blade (A) and leaf sheath (B) with their healthy tissues, and the ability of SDr/SDs to distinguish the leaf lesions from the leaf blade lesions (C) (source: [45]). SDr = red edge area; SDg = green peak area.

identification of sheath blight with an accuracy of more than 95%. Hyperspectral sensor could distinguish the spectral response curves of the diseased leaf blade (**Figure 7A**) and leaf sheath (**Figure 7B**) from their healthy tissues. Transformed data (SDr/SDg) could even distinguish the leaf blade lesions from the leaf sheath lesions (**Figure 7C**).

Infrared thermal sensors can detect radiation emitted in the thermal infrared range of 8 to 14 μm. They can be used to assess the surface temperature of leaves and plant canopies that are affected by water status [12]. Diseased plants and tissues usually suffer from water stress due to loss of healthy tissue, stomatal conductance, and photosynthesis, resulting in the changes in canopy temperature among different disease-related environments. Infrared thermal imaging has been used for the detection of more than six fungal and bacterial diseases in seven crops [10, 12]. Yamamoto et al. [46] reported the use of infrared thermal image to detect the occurrence of rice blast in Japan.

Using fluorescence technical approach for plant disease detection has not been widely studied as compared to other sensors described above. Fluorescence has been used for the detection of only several diseases, including wheat leaf rust [47] and citrus canker [48]. Fluorescence can assess the changes in photosystem II activity of plants under different levels of stress [12]. Infection causes chlorophyll degradation and reduced photosynthetic leaf area, resulting in the changes in the capacity of photosynthesis between diseased and healthy crops.

4.2 Remote sensing platforms

Satellite, airborne (aircraft), UAV, and ground are the most common platforms that have been used for the detection and monitoring of plant diseases. Applications of remote sensing on agriculture and disease detection and management are first studied and implemented using satellite and aircraft platforms equipped with remote sensors. For example, Colwell [49] conducted the first airborne imagery to monitor the occurrence of black stem rust of wheat and yellow dwarf of oat. Qin and Zhang [50] used Airborne Data Acquisition and Registration (ADAR) to detect and monitor the development of rice sheath blight with high accuracy for late development stages of the disease. Use of satellite image data has successfully detected and monitored rice diseases (sheath blight, blast, glume blight, and dwarf virus) [41, 51] and wheat diseases (powdery mildew and leaf rust) [52]. Ground-based sensors has also been used to detect crop diseases including fire blight of apple [53] and late blight of potato [54]. Satellite, airborne and ground-based remote sensing have been widely investigated and some of the developed techniques have been used for plant disease detection and management. However, most of the satellite and airborne remote sensing results and findings still remain in the research phases and wide implementations of theses remote sensing technologies for field crop disease detection and monitoring are limited due to their high costs of acquiring data, high technique demand for data processing, limited availability of needed and real-time data, and being inaccessible to end users (farmers). Ground-based remote sensing is difficult to meet the on-time detection of crop diseases at large scale farming setting for disease management. So far, these remote sensing technologies have not been used for the detection and monitoring of rice diseases in the US.

Recent advances in UAV remote sensing platform and data processing make it possible using remote sensor techniques to identify, detection and quantify plant diseases [12]. UAV-based remote sensing provides low costs, ease of use, high-resolution images, high efficiency, real-time inspection, and the ability to cover a large field scale. A study has been conducted by Garcia-Ruiz et al. [55] to compare the performance of UAV with a single engine fixed-wing aircraft using multispectral imaging sensor on the detection of citrus greening disease. The study found that UAV-based sensor provided 67 to 85% of identification accuracy whereas the accuracy was 61 to 74% with the aircraft-based sensor. The results of this comparative performance study demonstrate that UAV can be a low cost and reliable tool for crop disease detection.

Multi-rotors, helicopters, fixed wings, blimps, and flying wings are among the most UAVs used for crop phenotyping and disease detection [34]. Selection of UAVs is based on the purpose and budget of the research and implementation. Each UAV has its advantages and disadvantages in costs, flying ability (flying speed, altitude and duration), and payload capacity [34]. Multi-rotors are the most common used UAVs at present. Multi-rotor UAVs are low costs, can hover, and have low take-off and landing requirements. However, multi-rotor UAVs have their disadvantages of low payload, short flight duration, and being easy to be affected by weather conditions. Each UAV has a flight control system that can plan flight routes and setup flight parameters such as flight location, flight altitude, and flight speed.

UAV remote sensing provides an unprecedented spectral, spatial, and temporal resolution and an innovation tool for the detection of crop diseases [12, 56]. Investigations and reports on the use of UAV equipped with different sensors for field crop disease detection and monitoring have continued increase for the past five years [12]. In 2016 through 2019, there were at least 15 published research articles that involved more than 15 diseases in 12 field crops [12]. Sensors used in these studies include RGB sensor, multispectral sensor and infrared thermal sensor, and accuracy of disease identification and detection ranges from 0.64 to 0.97. These diseases studied are fire blight of apple [53], Ascochyta blight of chickpea [57], mistletoes of eucalyptus [58], myrtle rust of lemon myrtle [59], tar spot complex of maize [60], leaf spot of oilseed rape [61], myrtle rust of paperbark tea [62], late blight of potato [54], sheath blight [38] and NBLS of rice [41], needle blight of scots pine [63], gummy stem blight of watermelon [64], and leaf blotch, powdery mildew and yellow rust of wheat [65–67]. However, using UAV remote sensing for disease detection is still premature in comparison with research on the detection of crop drought stress [12]. Continuous research and further development on UAV remote sensing will make this innovation become a useful tool for farmers to detect and manage diseases. In this article, we focus on the development of UAV remote sensing for the detection and quantification of sheath blight and NBLS, two important rice diseases, as an example to demonstrate the usefulness of UAV remote sensing technology.

4.3 UAV remote sensing for sheath blight detection

A study was conducted in the field plots at Texas A&M AgriLife Research Center, Beaumont, TX, USA to evaluate the potential of using UAV remote sensing to detect and quantify sheath blight of rice [38]. A total of 67 rice cultivars and elite breeding lines with different levels of resistance to sheath blight were planted into plots of each consisting of seven 2.4-m rows, spaced 18 cm between rows. Each plot was divided into two equal-length sections, with one end section being inoculated with the sheath blight pathogen while the other end section left with no pathogen inoculation for the disease-free control. Ground truth sheath blight severity data were collected by visual assessment based on a scale of 0 to 9 where 0 represents no symptoms and 9 represents most severe in symptoms and damage.

Four-rotor Phantom 2 Vision+ UAV equipped with high resolution digit camera or multispectral camera was used to capture field plot images (**Figure 8**). The digit camera had 4384 x 3288 pixel resolution with three bands of red, green, and blue whereas the multispectral sensor Micansense RedEdge™ has five bands of blue, green, red, red edge, and near-infrared (**Figure 9**). The UAV flew at 27 m altitude to cover all 67 plots per image and 5.5 m to cover four plots per image, at a speed between 0 and 10 m/s depended on wind speed. Pix 4D was utilized to covert images automatically and Normalized Difference Vegetation Index (NDVI) and other vegetation indices of 67 plots were calculated and extracted by ArcGIS 9.1. Ground truth NDVI values of 67 plots were acquired by using GreenSeeker handheld crop sensor.

In this study, high-resolution 3-band RGB and 5-band multispectral images were analyzed to detect sheath blight-infected areas in the plots. Multispectral

Figure 8.
Four-rotor Phantom 2 Vision+ UAV equipped with a high-resolution digit RGB camera used in this study.

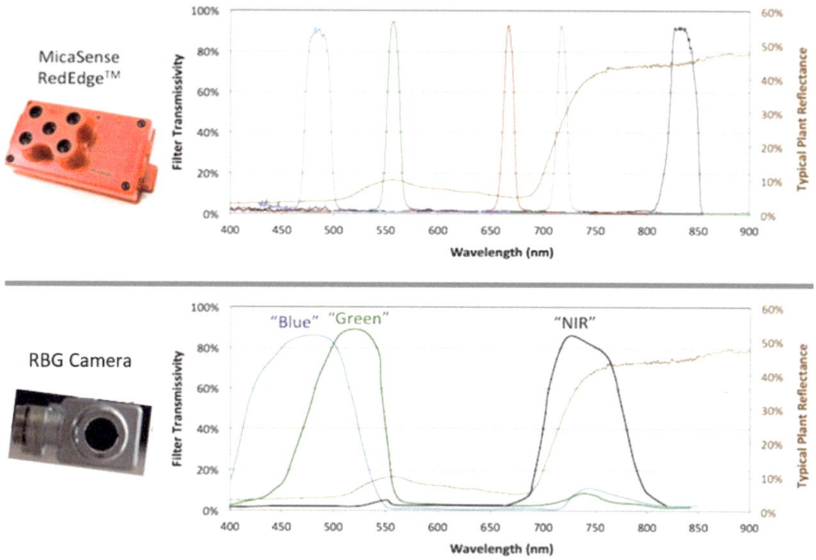

Figure 9.
Comparison of spectral responses of multispectral Micansense camera (upper) and digit RGB camera (lower) (source: [38]).

RBG image (**Figure 10B**) could more accurately reflect field environments such as green weeds, ground earth and plot shadow, and canopy characteristics including color, texture and structure information compared to original RGB image collected from digit camera (**Figure 10A**). Therefore, multispectral camera could provide more details than regular digit camera since the former has the narrow spectral band as shown in **Figure 9**. Transformation to HLS (hue, lightness, and saturation) (**Figure 10C**) from the false color image resulted in more apparent display of the sheath blight-infected areas with yellow to white in color in the plots. After NDVI values were calculated and the NDVIs map of 67 plots was developed (**Figure 10D**), it clearly showed the diseased areas were clearly differentiated from the healthy areas in each of the plots. The darker the image color, the more severe the sheath blight disease. These differentiation effects were more apparent compared to the differentiations made by original RGB, multispectral RBG and HLS images. This can be explained that red and near-infrared lights are more sensitive to the changes in canopy color from the healthy green color to the diseased yellow color and the changes in canopy structure from dense to sparse in density caused by sheath blight. Therefore, the vegetation index NDVI is a good indicator of different levels of sheath blight observed in this study.

Image-based NDVIs were also compared to the ground truth NDVIs acquired by GreenSeeker sensor and it was found that there was a strong correlation between them with a high R^2 value of 0.91 and a low RMSE value of 0.0854 (**Figure 11**). Imaged-based NBVIs were selected to determine their ability to quantify the levels of sheath blight. The results demonstrated that there was a good correlation between image-based NDVIs and ground truth sheath blight severity, with a R^2 value of 0.63 and a low RMSE value of 0.0852 (**Figure 12**).

The results of this study show multispectral image has more advantages of color and spectral information than regular RGB image, providing a strong ability to detect the sheath blight disease in the field. Use of multispectral camera can not only detect sheath blight but also quantify different levels of the disease in the field. An UAV equipped with a multispectral camera can be a new tool to aid in scouting and monitoring the development of sheath blight in rice.

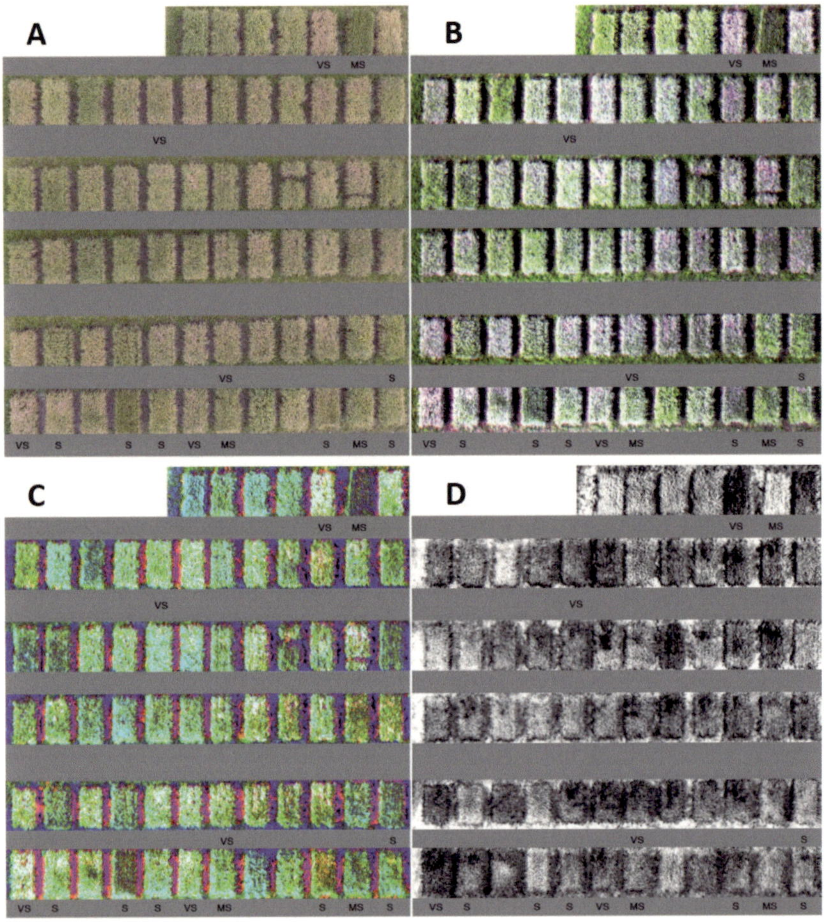

Figure 10.
Original RGB (A), multispectral RGB (B), HLS (C), and NDVI (D) images of 67 field plots, with rice cultivars and elite breeding lines having different levels of resistance to sheath blight, at Beaumont, Texas, USA (source: [38]). VS = very susceptible; S = susceptible, and MS = moderately susceptible.

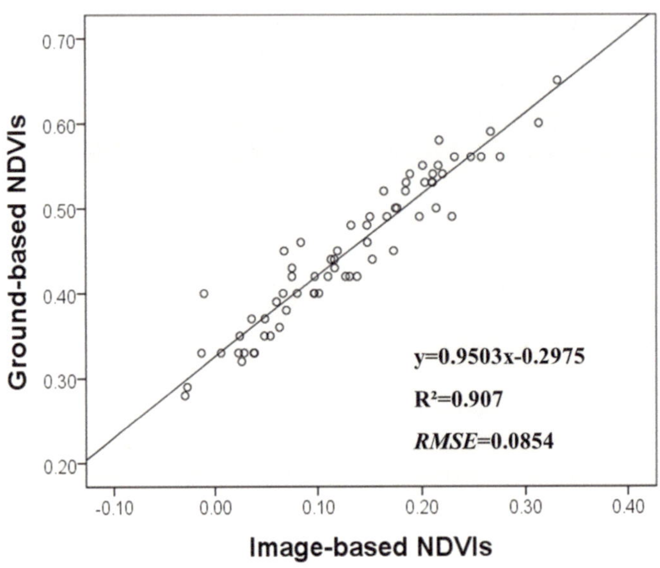

Figure 11.
Correlation between image-based NDVIs and ground-based NDVIs (source: [38]).

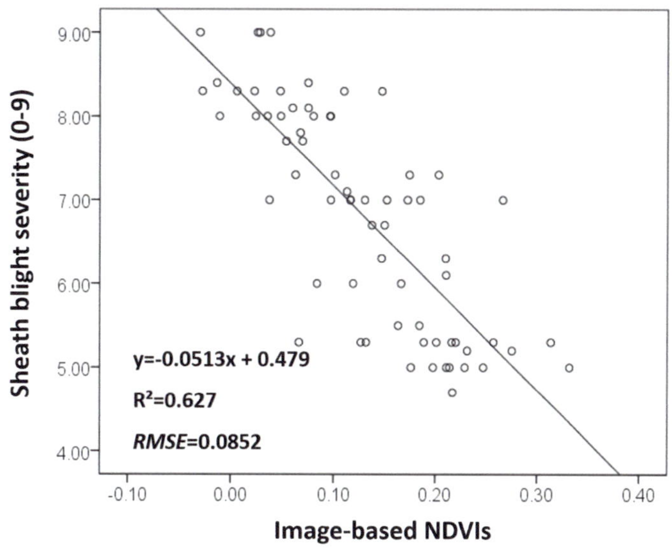

Figure 12.
Correlation between image-based NDVIs and ground truth sheath blight severity (source: [38]).

4.4 UAV remote sensing for NBLS detection

The spread of NBLS in the field is different from the spread of sheath blight. NBLS spreads by air-borne spores and its distribution in the field is uniform whereas sheath blight is soilborne and it spreads in the field in an aggregated pattern. A study was conducted at Texas A&M AgriLife Research Center, Beaumont, TX, USA to evaluate the performance of UAV remote sensing on the detection and quantification of the NBLS disease in field research plots [41]. Rice cultivar Presidio, susceptible to NBLS, was seed drilled in 40 plots (**Figure 13**). Plots consisted of seven 4.9-m rows, spaced 18 cm between rows, with field blocks separated by 2.7-m wide allies. NBLS developed from natural inoculum and the symptoms initially appeared at the tillering stage and developed progressively with time, reaching high levels of disease severity as rice approached maturity. Differentiation in NBLS severity among the 40 plots was achieved by applying with 10 different

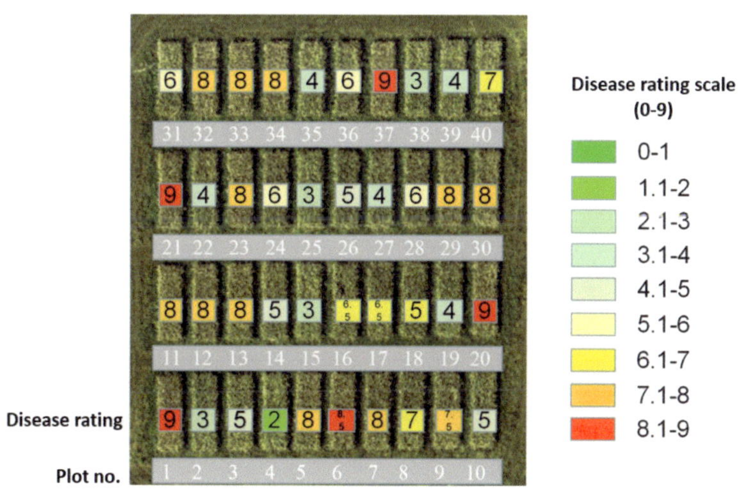

Figure 13.
Field plots and narrow brown leaf spot (NBLS) severity ratings at Beaumont, Texas, USA (source: [41]).

fungicide treatments at the mid-boot stage. NBLS severity was rated by visual symptoms on a scale of 0 to 9 where 0 represents no symptoms and 9 represents most severe in symptoms and damage (leaves dead) (**Figure 13**).

Four-rotor DJI INSPIRE 2 UAV equipped with Sentera Multispectral Double 4 K sensor was used to capture field plot images (**Figure 14**). The multispectral camera offers five spectral bands of blue, green, red, red edge, and near-infrared (NIR) and can capture 12.3 MP still images. UAV images were acquired by flying the UAV over the field plots at the altitudes of 10 and 15 m, with a ground sampling resolution of 0.42 and 0.63 cm, respectively. Image data were preprocessed for image mosaic, radiation correction, and band coincidence using Photoscan 1.4.1 and Envi

Figure 14.
Four-rotor DJI INSPIRE 2 UAV equipped with Sentera Multispectral Double 4K camera used in this study.

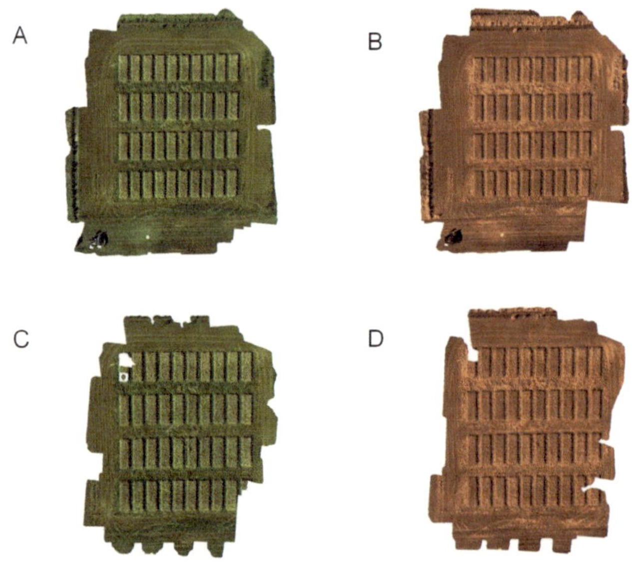

Figure 15.
RGB (A) and NIR (B) images at the 10-m flight altitude, and RGB (C) and NIR (D) images at the 15-m altitude (source: [41]).

5.3 software. Different vegetation indices (Vis) and color space HIS, HSV, HSL and YCbCr were extracted from the ENVI mass cut images and used to determine their performance on the detection of different levels of NBLS severity at the two flight altitudes. Most effective vegetation index and color space were selected and the inversion model of NBLS with good correlation was developed to predict the levels of NBLS severity.

Results of comparison of the correlations between Vis or color features calculated from RGB image and NIR image (**Figure 15**) and NBLS severity indicate that using RGB image was more suitable for the assessment of NBLS than NIR image and that Excess Green minus Excess Red (EXGR) at the 15-m flight altitude was the best vegetation index to detect and quantify the different levels of NBLS severity (R^2 = 0.89, *RMSE* = 0.70). This vegetation index offers more spectral information than other vegetation indexes and thus it is more effective to detect the NBLS disease. It is also found that the EXGR is more effective to detect high levels of the NBLS disease. The findings from this study demonstrate that it is feasible to use UAV multispectral sensor to detect and assess the levels of NBLS in the field.

5. UAV used as a tool for disease management

Disease detection, decision making, and control action are the three essential steps for effective management of crop diseases. UAV remote sensing itself cannot directly serve for control of plant disease. However, it provides an innovative tool for disease assessment. Effective disease assessment can ensure to make a correct decision that results in employing a proper control measure for control of a crop disease. Fungicide application is one of the most effective control measures for disease management. With the development of UAV technology, the use of UAV for aerial fungicide application has become a new means for control of diseases in rice and other crops in recent years [15].

Since the first UAV used for aerial insecticide applications to control insect pests in rice, soybean and wheat in Japan in 1990 [68], UAV-based aerial spraying technology has developed quickly in agricultural aviation applications, especially in the last five years [15]. UAV aerial application has several advantages over traditional aircraft aerial spraying and ground application. UAV sprayer can operate at low altitude and suspend in the air to achieve high-precision positioning with GPS [15], which can reduce pesticide draft potential and the amount of pesticides used [18]. Its downward airflow generated by rotors can help pesticide droplets penetrate dense canopies to improve application efficacy. UAV sprayer can operate on high crops and in areas with steep or mountainous terrains and can cover a large area whereas ground application is unable to do so. UAV aerial spraying is lower costs and more flexible in operation than typical aircraft aerial application. Because of these advantages, research and use of UAV aerial spraying technology have been increased quickly in recent years in many countries, especially China and Japan. In 2016, there were 4,262 UAV sprayers in operation and more than 476,000 hectares of field crops, including rice, treated with UAV aerial sprays in China [15]. In the 2020 statistics, there were approximately 170 types of pesticide application UAVs, and 55,000 UAVs flown; treated crop areas were more than 450 million hectares in China [16]. In Japan, use of UAV is widespread and more than 2,000 UAV applicators are in operation to spray approximately 40% of the rice acreage [17, 18].

Helicopter and quadcopter are among the most common UAVs applicators used to spray fungicides, insecticides, and herbicides in various field crops [69, 70]. Tank capacity and duration of flight are two key technical parameters of UAVs. Although they vary with UAV, most UAVs have the tank capacity of 16 to 20 litters and the duration of flight of less than 30 minutes at present [16]. Flight altitude usually ranges from 2 to 10 m depended on individual UAVs [69, 70]. Spraying swath width can range from 3 to 15 m, and work efficiency can be anywhere from 0.7 to 13 hectares per hour. Optimization of these operation parameters are important to improve spraying efficacy. Research has been conducted to determine flight altitude, flight speed, and spraying swath width that are more suitable for various UAVs. For example, using WPH642 helicopter to spray on rice, the best flight altitude was found to be 2 m and the best flight speed was 1.5 m per second whereas for P20 quad-rotor UAV, the best flight altitude was 2 m and the best flight speed was 3.7 m per second [69]. Selection of optimal operation parameters can achieve optimum spray droplet deposition into canopies to improve spraying efficacy. Flight speed, flight altitude, and nozzle flow rate are three factors in order of importance that affect droplet deposition distribution [69].

UAV sprayers have the advantages of low equipment costs, low fuel consumption, low spraying volumes, no pesticide contamination risks to operators, and high productivity compared to traditional big aircrafts and ground tractors [18]. With recent advances in UAV development and communication technology, research and use of UAVs for applications of fungicides and other pesticides have been significantly increased in field crops, including rice, in China, Japan, India and many other countries. However, it is expected that UAV aerial spraying cannot replace more conventional means such as aircraft aerial application. Among other disadvantages, UAV sprayers are small in tank capacity and cannot load large amount of pesticides for large farms, especially rice farms in the US with an average of 243 hectares per farm [33]. UAV applications have significantly short duration of flight and cannot cover large spray area a time. On the other hand, UAV sprayers can be used in situations where they can be advantageous. UAV spraying technology can be incorporated into current crop production systems for precision fungicide applications [56]. Precision applications can be more effective to control diseases that develop in cluster patterns like rice sheath blight. In the US, application of UAV remote sensing for crop disease detection and the use of UAV for pesticide application in rice and other field crops are still in its infancy. The US falls behind other countries, especially China, in research and use of UAV remote sensing in plant disease detection and management. Continued research and more monetary investment are needed in research and adopting this new technology to keep up with other countries like China and Japan, which is worthy as much as $10 billion US dollars a year in productivity [17].

6. Conclusion and prospects

Quick and accurate diagnosis and detection of rice diseases, especially sheath blight, are the first essential step for effective disease management to reduce production costs and maximize production returns. However, traditional disease detection methods based on visual symptoms are time-consuming and laborious, and its accuracy is highly dependent on the knowledge and experience of the inspector. UAV remote sensing provides an unprecedented spectral, spatial, and temporal resolution that can distinguish diseased tissue, plant and cropped area from healthy tissue, plant and cropped area based on the characteristics of disease symptoms. Sheath blight and NBLS have their own characteristics in symptoms and disease development pattern. Among the five remote sensors commonly used for assessing abiotic and biotic stresses of crops, RGB sensor, multispectral sensor, and hyperspectral sensor have been successfully used to detect sheath blight; multispectral sensor has been used to detect and quantify NBLS; and infrared thermal sensor can be used to detect the occurrence of rice blast. So far, there have been no reports on the use of fluorescence imaging for rice disease detection. Multi-rotors, helicopters, fixed wings, blimps, and flying wings are among the most UAVs used for crop phenotyping and disease detection. Selection of a suitable unmanned aerial system is important to acquire best imaging data that can be processed and modeled for the detection and quantification of crop diseases. Each UAV has its advantages and disadvantages in costs, flying ability, and payload capacity; each sensor has its own advantages and limitations in acquiring spectral information. In addition, UAV can also be used as an innovative aerial fungicide applicator for disease control. UAV sprayers can operate at low altitudes, fly with various speeds, and apply with low fungicide volumes, making it more suitable for situations where precision fungicide applications are needed for more economically and environmentally effective control of diseases such as rice sheath blight with cluster occurrence. The use of UAV for pesticide application has become a new disease control practice in rice and other crops in many countries, especially China and Japan.

However, applications of UAV technology on disease detection and fungicide application are still in the early stages of development. There remain many technical limitations and application challenges in the research and development of these technologies. Current UAV systems have limited battery capacity, tank capacity and payload. Sensors are usually expensive. There is lack of supporting technologies for UAV-based aerial spraying, such as optimization of nozzle-related canopy deposition, and the formulations of pesticide materials and adjuvants specific for UAV spraying. Although the UAV industry growth is very quickly in recent years, there is apparent lack of standard design of UAVs (rotor designs, types of engines, tank sizes, and nozzles types), which creates challenges for agrichemical manufacturers to develop recommended guidelines for product use. There needs the development and improvement of methods for big image data processing and disease detection model establishment. Early disease detection is critical for timely fungicide application for effective disease control. However, most UAV remote sensing methods reported in the literature are less effective for the detection of diseases at the early stages, such as rice sheath blight and NBLS described in this article. Current strict airspace regulations enforced for UAV operations in most countries, especially the US, limit the research and development of UAV-related technologies. Therefore, the progress on the adoption and commercialization of UAV technologies depends on collaborative research between agronomists and engineers, effective education and extension, partnerships between agricultural UAV manufactures and chemical manufacturers, and effective airspace regulations for UAVs. With improving performance of UAVs on flight duration and payload, reduced costs of sensors, and the development and improvement of methods for big image data processing and models for disease detection and monitoring, it is expected that UAV remote sensing will become an effective tool widely used for the detection of diseases in rice and other crops and that UAV spraying technology can become a new means for control of many crop diseases in situations where traditional aircraft aerial spraying and ground spraying are unavailable or infeasible. Along with research breakthroughs of digitalization and artificial intelligence for precision application across fertilizer, pest, and crop management needs, this innovative UAV technology will become a core tool in a farmer's precision equipment mix in the future.

Author details

Xin-Gen Zhou[1]*, Dongyan Zhang[2] and Fenfang Lin[3]

1 Texas A&M AgriLife Research and Extension Center, Texas A&M University System, Beaumont, Texas, USA

2 Anhui Engineering Laboratory of Agro-Ecological Big Data, Anhui University, Hefei, Anhui, China

3 School of Remote Sensing and Geomatics Engineering, Nanjing University of Information Science and Technology, Nanjing, China

*Address all correspondence to: xzhou@aesrg.tamu.edu

IntechOpen

© 2021 The Author(s). Licensee IntechOpen. This chapter is distributed under the terms of the Creative Commons Attribution License (http://creativecommons.org/licenses/by/3.0), which permits unrestricted use, distribution, and reproduction in any medium, provided the original work is properly cited.

References

[1] World Agriculture Production. 2020. World rice production 2020/2021. Available from http://www.worldagriculturalproduction.com/crops/rice.aspx

[2] Singh V, Zhou XG, Ganie Z, Valverde B, Avila L, Marchesan E, Merotto A, Zorrilla G, Burgo N, Norsworthy J, Bagavathianna, M. Rice Production in the Americas. In: Chauhan BS, Jabran K, Mahajan G, Editors. Rice Production Worldwide. Springer International Publishing; 2017. p. 137-168

[3] Allen TW, Growth DE, Wamishe YA, Espino L, Jones G, Zhou XG. 2020. Disease loss estimates from the rice producing states in the United States: 2018 and 2019. In: Proc. Rice Tech. Wrkg Grp; 24-27 February; Orange Beach, Alabama; 2020.

[4] Groth DE. Azoxystrobin rate and timing effects on rice sheath blight incidence and severity and rice grain and milling yields. 2005; Plant Dis. 89:1171-1174

[5] Uppala S, Zhou XG. Field efficacy of fungicides for management of sheath blight and narrow brown leaf spot of rice. 2018; Crop Prot. 104:72-77

[6] Marchetti MA, Bollich CN. Quantification of the relationship between sheath blight severity and yield loss in rice. 1991; Plant Dis. 75:773-775

[7] Hollier C. Narrow brown leaf spot. In: Webster RK, Gunnell PS, editors. Compendium of Rice Diseases. APS Press, Minnesota; 1992. p. 18

[8] Zhou XG, Jo YK. Disease management. In: Way MO, McCauley GM, Zhou XG, Wilson LT, Brandy M. editors. 2014 Texas Rice Production Guidelines. Texas AgriLife Research, and Texas Rice Research Foundation; 2014. p. 44-57. Available from: https://beaumont.tamu.edu/eLibrary/Bulletins/2014_Rice_Production_Guidelines.pdf

[9] Bauer ME. Remote sensing as a means of detecting crop disease. 1971; LARS Technical Reports. Paper 1. DOI: http://docs.lib.purdue.edu/larstech/1

[10] Gogoi NK, Deka B, Bora LC. Remote sensing and its use in detection and monitoring plant diseases: A review. 2018; Agricultural Reviews 39:307-313

[11] Sahoo RN, Ray SS, Manjunath KR. Hyperspectral remote sensing of agriculture. 2015; Current Science 108:848-859

[12] Oerke EC. Remote sensing of diseases. 2020; Ann. Rev. Phytopathol. 58:225-252. DOI: https://doi.org/10.1146/annurev-phyto-010820-012832

[13] Kuska M, Wahabzada M, Leucker M, Dehne HW, Kersting K, Oerke EC, Steiner U, Mahlein A-K. Hyperspectral phenotyping on the microscopic scale: Towards automated characterization of plant pathogen interactions. 2015; Plant Methods 11:28. DOI: https://doi.org/10.1186/s13007-015-0073-7

[14] Bauriegel E, Herppich W. Hyperspectral and chlorophyll fluorescence imaging for early detection of plant diseases, with special reference to Fusarium species infections on wheat. 2014; Agriculture 4:32-57

[15] Lan YB, Chen SD. Current status and trends of plant protection UAV and its spraying technology in China. 2018; Int. J. Precis Agric. Aviat. 1:1-9

[16] Gallup C, Barbosa R. Rapid adoption of drone application technology in Asia-Pacific: A story of grower demands

defining new management techniques. In: APS Workshop of Unmanned Aerial Vehicles for making plants healthy- Do we have a winner? 3 December 2020; Minneapolis, USA.

[17] Heller M. Could drones spray crops? EPA is considering it. 2020; E&ENews. Available from https://www.eenews.net/stories/1063300501

[18] Anand K, Goutam R. An Autonomous UAV for pesticide spraying. 2019; Int. J. of Trend in Sci. Research and Development 3: 986-990. DOI: https://www.ijtsrd.com/papers/ijtsrd23161.pdf

[19] Rush MC, Lee FN. Sheath blight. In: Webster RK, Gunnell PS, editors. Compendium of Rice Diseases. APS Press, St. Paul, MN; 1992. p. 22-23.

[20] Uppala S, Zhou XG. Rice sheath blight. 2018; The Plant Health Instructor. DOI:10.1094/PHI-I-2018-0403-01

[21] Liu G, Zhou XG. Narrow brown leaf spot and its management. 2011; Texas Rice 11: 1-7

[22] Uppala S, Zhou XG. Optimum timing of propiconazole to manage narrow brown leaf spot in the min and ratoon crops in Texas. 2019; Crop Prot. 124:104854. DOI: https://doi.org/10.1016/j.cropro.2019.104854

[23] Matsumoto M, Furuya N, Takanami Y, Matsuyama N. Rapid detection of *Rhizoctonia* species, causal agents of rice sheath diseases, by PCR-RFLP analysis using an alkaline DNA extraction method. 1997; Mycoscience 38:451-454

[24] Webster RK, Gunnell PS, editors. Compendium of Rice Diseases. APS Press, St. Paul, MN; 1992. 62 p.

[25] Cartwright R, Lee F. Management of rice diseases. In: Slaton N, editor. Rice Production Handbook. University of Arkansas Cooperative Extension Service, Little Rock, AR. MP1925; 2000. p. 87-100

[26] Groth D, Hollier C, Rush C. Disease management. In: Saichuk J, editor. Louisiana Rice Production Handbook. LSU AgCenter. Pub. 2321; 2009. p. 72-92

[27] Espino L. Fungicides to control rice diseases in California. In: Proc. Rice Tech. Wrkg Grp; 24-27 February; Orange Beach, Alabama; 2020.

[28] Johanson A, Turner HC, McKay GJ, Brown AE. A PCR-based method to distinguish fungi of the rice sheath-blight complex, *Rhizoctonia solani*, *R. oryzae* and *R. oryzae-sativae*. 1998; FEMS Microbiol. Lett. 162:289-294

[29] Sayler RJ, Yang Y. Detection and quantification of *Rhizoctonia solani* AG-1 IA, the rice sheath blight pathogen, in rice using real-time PCR. 2007; Plant Dis. 91:1663-1668

[30] Zhou XG, Tabien RE, Way MO. First report of white leaf streak of rice caused by *Mycovellosiella oryzae* in Texas. 2010; Plant Dis. 94:639-639. DOI: https://doi.org/10.1094/PDIS-94-5-0639B

[31] Shahjahan AKM, Rush MC, Jones JP, Groth DE. First report of the occurrence of white leaf streak in Louisiana rice. 1998; Plant Dis. 82:1282-1282.

[32] Gunnell PS. White leaf streak. In: Webster R K, Gunnell P S editors. Compendium of Rice Diseases. APS Press, St. Paul, MN; 1992. p. 20

[33] Childs N, Skorbiansky SR, McBridge WD. U.S. rice production changed significantly in the new millennium but remained profitable. USDA ERS. May 4, 2020. Available from https://www.ers.usda.gov/amber-waves/2020/may/us-rice-production-changed-significantly-in-the-new-millennium-but-remained-profitable/

[34] Yang G, Liu J, Zhao C, Li Z, Huang Y, Yu H, Xu B, Yang X, Zhu D, Zhang X, Zhang R, Feng H, Zhao X, Li Z, Li H, Yang H. Unmanned aerial vehicle remote sensing for field-based crop phenotyping: current status and perspectives. 2017; Front. Plant Sci. 8:1111. DOI: 10.3389/fpls.2017.01111

[35] Camargo A, Smith JS. Image pattern classification for the identification of disease causing agents in plants. 2009; Comput. Electron. Agric. 66:121-125

[36] Bock CH, Parker PE, Cook AZ, Gottwald TR. Visual rating and the use of image analysis for assessing different symptoms of citrus canker on grapefruit leaves. 2008; Plant Dis. 92:530-541

[37] Neumann M, Hallau L, Klatt B, Kersting K, Bauckhage C. (2014). Erosion band features for cell phone image based plant disease classification. In: Proceedings of the 22nd International Conference on Pattern Recognition (ICPR); 24-28 August 2014; Stockholm, Sweden; 2014. p. 3315-3320

[38] Zhang D, Zhou XG, Zhang J, Lan Y, Xu C, Liang D. Detection of rice sheath blight using an unmanned aerial system with high-resolution color and multispectral imaging. 2018; PLoS ONE 13: e0187470. doi:10.1371/journal.pone.0187470

[39] Kurniawati NN, Abdullah SNHS, Abudullah S, Abdullah S. Investigation on image processing techniques for diagnosing paddy diseases. 2009; IEEE Xplore 11069264. DOI: 10.1109/SoCPaR.2009.62

[40] Lu Y, Yi SJ, Zeng NY, Liu YR, Zhang Y. Identification of rice diseases using deep convolutional neural networks. 2017; Neurocomputing 267:378-384

[41] Cai N, Zhou XG, Yang Y, Wang J, Zhang D, Hu R. Use of UAV images to assess narrow brown leaf spot severity in rice. 2019; Int. J. Precis. Agric Aviat. 2:38-42. DOI: 10.33440/j.ijpaa.20190202.47

[42] Shi Y, Huang W, Ye H, Ruan C, Xing N, Geng Y, Dong Y, Peng D. Partial least square discriminant analysis based on normalized two-stage vegetation indices for mapping damage from rice diseases using PlanetScope datasets. 2018; Sensors 18:1901. DOI: 10.3390/s18061901

[43] Kobayashi T, Kanda E, Kitada K, Ishiguro K, Torigoe Y. Detection of rice panicle blast with multispectral radiometer and the potential of using airborne multispectral scanners. 2001; Phytopathology 91:316-323

[44] Feng L, Wu B, Zhu S, Wang J, Su Z, Liu F, He Y, Zhang C. Investigation on data fusion of multisource spectral data for rice leaf diseases identification using machine learning methods. 2020; Front. Plant Sci. 11:577063. DOI: 10.3389/fpls.2020.577063

[45] Lin F, Guo S, Tan C, Zhou XG, Zhang D. Identification of rice sheath blight through spectral responses using hyperspectral images. 2020; Sensors 20:6243. DOI:10.3390/s20216243

[46] Yamamoto H, Suzuki Y, Iwano M, Hayakawa S. Remote sensing of occurrence place of rice blast disease by infrared thermal image. 1995; Jpn. J. Crop Sci. 64:467-474

[47] Tischler YK, Thiessen E, Hartung E. Early optical detection of infection with brown rust in winter wheat by chlorophyll fluorescence excitation spectra. 2018; Comput Electron. Agric. 146:77-85

[48] Belasque L, Gasparoto MCG, Marcassa LG. Detection of mechanical and disease stresses in citrus plants by fluorescence spectroscopy. 2008; Applied Optics 47:1922-1926

[49] Colwell RN. Determining the prevalence of certain cereal crop diseases by means of aerial photography. 1956; Hilgardia 26:223-286

[50] Qin Z, Zhang M. Detection of rice sheath blight for in-season disease management using multispectral remote sensing. 2005; Int. J. Appl. Obs. 7:115-128

[51] Ghobadifar F, Wayayok A, Mansor S, Shafri HM. Detection of BPH (brown planthopper) sheath blight in rice farming using multispectral remote sensing. 2016; Geomatics, Nat. Hazards Risk 7:237-247. DOI: http://dx.doi.org/10.1080/19475705.2014.885468

[52] Franke J, Menz G. Multi-temporal wheat disease detection by multi-spectral remote sensing. 2007; Precis. Agric. 8:161-172

[53] Jarolmasjed S, Sankaran S, Marzougui A, Kostick S, Si Y, Vargas JJ, Evans K. High-throughput phenotyping of fire blight disease symptoms using sensing techniques in apple. 2019; Front. Plant Sci. 10:576

[54] Franceschini MHD, Bartholomeus H, van Apeldoorn DF, Suomalainen J, Kooistra L. Feasibility of unmanned aerial vehicle optical imagery for early detection and severity assessment of late blight in potato. 2019; Remote Sens. 11:224. DOI: https://doi.org/10.3390/rs11030224

[55] Garcia-Ruiz F, Sankaran S, Maja JM, Lee WS, Rasmussen J, Ehsani R. Comparison of two aerial imaging platforms for identification of Huanglongbing-infected citrus trees. 2013; Comput. Electron. Agric. 91:106-115

[56] Maes WH, Steppe K. Perspectives for remote sensing with unmanned aerial vehicles in precision agriculture. 2019; Trends Plant Sci. 24:152-164

[57] Zhang CY, Chen WD, Sankaran S. High-throughput field phenotyping of Ascochyta blight disease severity in chickpea. 2019; Crop Protec. 125:104885. DOI: 10.1016/j.cropro.2019.104885

[58] Maes WH, Huete AR, Avino M, Boer MM, Dehaan R, Pendall E, Griebel A. Steppe K. Can UAV-based infrared thermography be used to study plant-parasite interactions between mistletoe and eucalypt trees? 2018; Remote Sens.10: 2062

[59] Heim RHJ, Wright IJ, Carnegie AJ, Taylor D, Oldeland J. Multispectral, aerial disease detection for myrtle rust (*Austropuccinia psidii*) on a lemon myrtle plantation. 2019; Drones 3:25. DOI: 10.3390/drones3010025

[60] Loladze A, Rodrigues FA, Toledo F, San Vicente F, Gérard B, Boddupalli MP. Application of remote sensing for phenotyping tar spot complex resistance in maize. 2019; Front. Plant Sci. 10:552. DOI:10.3389/fpls.2019.00552

[61] Cao F, Liu F, Guo H, Kong WW, Zhang C, He Y. Fast detection of Sclerotinia sclerotiorum on oilseed rape leaves using low-altitude remote sensing technology. 2018; Sensors 18:4464. DOI: https://doi.org/10.3390/s18124464

[62] Sandino J, Pegg G, Gonzalez F, Smith G. Aerial mapping of forests affected by pathogens using UAVs, hyperspectral sensors, and artificial intelligence. 2018; Sensors 18:944

[63] Smigaj M, Gaulton R, Suarez JC, Barr SL. Canopy temperature from an Unmanned Aerial Vehicle as an indicator of tree stress associated with red band needle blight severity. 2019; Forest Ecol. Manag. 433: 699-708

[64] Kalischuk M, Paret ML, Freeman JH, Raj D, Silva SD, Eubanks S, Wiggins DJ, Lollar M, Marois JJ, Mellinger HC, Das J. An

improved crop scouting technique incorporating unmanned aerial vehicle-assisted multispectral crop imaging into conventional scouting practice for gummy stem blight in watermelon. 2019; Plant Dis. 103:1642-1650.

[65] Huang YB, Thomson SJ, Hoffmann WC, Lan YB, Fritz BK. Development and prospect of unmanned aerial vehicle technologies for agricultural production management. 2013; Int. J. Agric. & Biol. Eng. 6: 1-10

[66] Liu W, Cao X, Fan J, Wang Z, Yan Z, Luo Y, West JS, Xu X, Zhou Y. Detecting wheat powdery mildew and predicting grain yield using unmanned aerial photography. 2018; Plant Dis. 102:1981-1988

[67] Su J, Liu C, Coombes M, Hu X, Wang C, Xu X, Li Q, Guo L, Chen WH. Wheat yellow rust monitoring by learning from multispectral UAV aerial imagery. 2018; Comput. Electron. Agric. 155:157-166

[68] Johnson LF, Bosch DF, Williams DC, Lobitz BM. Remote sensing of vineyard management zones: implications for wine quality. 2001; Appl. Eng. Agric. 17: 557-560

[69] Lan YB, Chen S, Fritz BK. Current status and future trends of precision agricultural aviation technologies. 2017; Int. J. Agric & Biol Eng. 10:1-17

[70] Kim J, Kim S, Ju C, Son HI. Unmanned aerial vehicles in agriculture: a review of perspective of platform, control, and applications. 2019; IEEE Access 7:105100-105115. DOI: 10.1109/ACCESS.2019.2932119

Chapter 4

Blister Blight Disease of Tea: An Enigma

Chayanika Chaliha and Eeshan Kalita

Abstract

Tea is one of the most popular beverages consumed across the world and is also considered a major cash crop in countries with a moderately hot and humid climate. Tea is produced from the leaves of woody, perennial, and monoculture crop tea plants. The tea leaves being the source of production the foliar diseases which may be caused by a variety of bacteria, fungi, and other pests have serious impacts on production. The blister blight disease is one such serious foliar tea disease caused by the obligate biotrophic fungus *Exobasidium vexans*. *E. vexans*, belonging to the phylum basidiomycete primarily infects the young succulent harvestable tea leaves and results in ~40% yield crop loss. It reportedly alters the critical biochemical characteristics of tea such as catechin, flavonoid, phenol, as well as the aroma in severely affected plants. The disease is managed, so far, by administering high doses of copper-based chemical fungicides. Although alternate approaches such as the use of biocontrol agents, biotic and abiotic elicitors for inducing systemic acquired resistance, and transgenic resistant varieties have been tested, they are far from being adopted worldwide. As the research on blister blight disease is chiefly focussed towards the evaluation of defense responses in tea plants, during infection very little is yet known about the pathogenesis and the factors contributing to the disease. The purpose of this chapter is to explore blister blight disease and to highlight the current challenges involved in understanding the pathogen and pathogenic mechanism that could significantly contribute to better disease management.

Keywords: tea, blister blight, *Exobasidium vexans*, Basidiospore, defense, control

1. Introduction

Tea is one of the most popular beverages worldwide, having gained popularity for its taste, stimulating effect, various medicinal properties, and related health benefits. Tea is processed from the leaves of evergreen, woody, and perennial tea plants (*Camellia sinensis*) belonging to the family Theaceae. Three indigenous varieties of tea plant are found *viz. C. sinensis* (L.) O. Kuntze (China type), *C. assamica* (Assam type), and (3) *C. assamica* sub *spp lasiocalyx* (Planchon ex Watt.) or Cambod type. These varieties are capable of cross-pollination and interbreeding, resulting in heterogenous hybrids. Under natural conditions, the tea plants can grow up to a maximum height of 15 m, while cultivated tea plantations are maintained as a bush with a height of 60–100 cm, which facilitates the plucking of tender leaves [1]. Tea was first used as a beverage in China in 2737 B.C and was introduced in India for commercial production by the erstwhile colonial British, through the East India Company in 1853 [2]. Tea plantations in India are found in three main geographic regions - the Northeast, (Assam, West Bengal, Tripura, Sikkim, Manipur, Nagaland, Meghalaya, Arunachal

IntechOpen

Pradesh, and Mizoram), the South (Karnataka and Tamil Nadu), and the Northwest (Himachal Pradesh and Uttarakhand) [3]. The Darjeeling Tea from northeast India has been signified as the world's premium, and exotically flavored tea owing to its unique flavor and aroma, earning itself a GI tag.

Presently tea is cultivated worldwide across 61 countries of which China, India, Kenya, Sri Lanka, and Vietnam are the largest tea-producing countries, contributing 77% of world production and 80% of global exports (**Figures 1** and **2**). China is reported to produce 2700 million kg of tea of which 366.6 million Kg were exported for the year 2019 with one million hectares under tea cultivation. This was followed by India and Kenya with 1390.1 and 458.9 million kg of tea production. Recently, Kenya was listed with the highest exporter of tea for the year 2019 with 392.6 million kg of exports. India accounts for 23% of the total world tea production with an

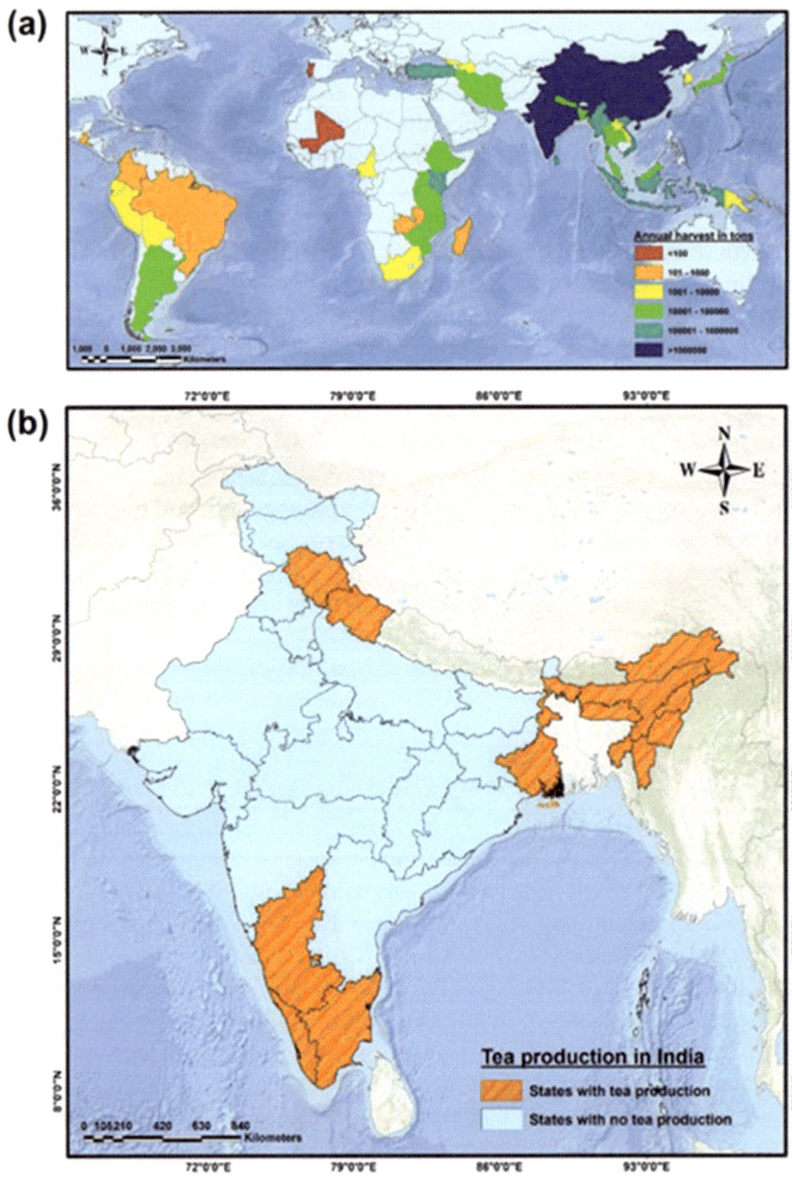

Figure 1.
Tea producing countries worldwide (a), area of tea cultivation in India (b).

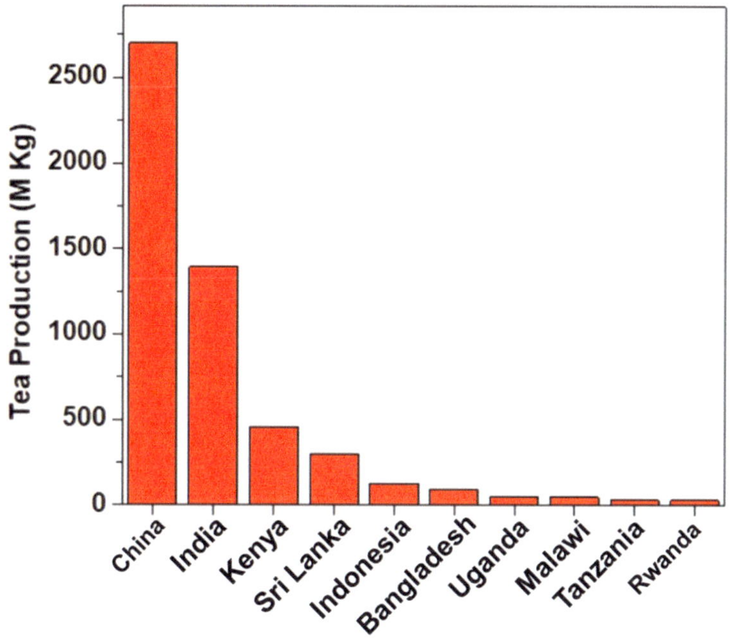

Figure 2.
Tea production of top 10 countries for the year 2019–2020 (source: Tea board of India).

area of 400,000 hectares under tea cultivation. This contributes to about US $803 million to the Indian economy **(Tea board of India)**. India is followed by Sri Lanka with an export worth US $721.3 million. In addition to contributing majorly to the economies of the tea growing countries, the tea industry also provides livelihood to a significant part of the population in these countries.

Different forms of tea have been produced from the same tea plantations depending on the method of processing and plucking of leaves. Some of them are black tea, green tea, oolong tea, white tea, pure tea, and dark tea. Black tea is the most popular form of tea produced in all the major tea producing countries including India, Kenya, and Sri Lanka. Production of black tea in India accounts for 85% of total worldwide production and green tea is being produced by a few tea gardens. Green tea is the most popular form of tea in China followed by black tea.

Tea being a perennial and monoculture crop, the microclimate of tea plantations makes it prone to various pests and pathogens [4]. Chen and Chen recorded around 400 pathogens 507 species of fungi infecting tea plants [5, 6]. Although all the parts of the tea plant including leaf, stem, and root are prone to infection, the pathogens invading leaf parts are of great concern as the main source of commercial production of tea is the young and fresh leaves. The incidence of diseases in the leaves significantly affects the crop yield and quality of made tea. This also directly affects the economy of agronomic countries where tea is considered an important cash crop. The various diseases of tea can be categorized into primary and secondary diseases. In case of the primary diseases, the pathogens directly invade healthy tea bushes while secondary diseases are caused by weaker parasites infecting already diseased/infected tea bushes. In this context, some of the most important diseases infecting the leaf, stem, and root of tea plantations are listed in **Table 1**.

Blister blight disease is one of the most serious primary foliar tea diseases that significantly affects the crop yield and quality throughout various regions of tea-producing countries across the world. The causal organism of Blister blight disease is the biotrophic fungus *Exobasidium vexans* Massee. Peal was the first to recognize

Infection Site	Disease	Causal organism	References
Leaf	Blister blight	*Exobasidium Vexans*	[7, 8]
	Black rot	*Corticium invisum* and *Corticium theae*	
	Leaf red rust	*Cephaleorus mycoidae*	
	Brown blight	*Colletotrichum camelliae*	
	Gray blight	*Pestalozzia Theae*	
Stem	Poria Disease	*Poria hypobrunnea*	
	Nectria	*Nectria cinnabarina*	
	Black root rot	*Rosellinia arcuata*	
	Brown root rot	*Fomes lamaoensis*	
	Jew's ear fungus	*Auricularia auricula*	
	Thorny blight	*Aglaospora aculeata*	
	Ganoderma	*Ganoderma applanatum* and *Ganoderma lucidum*	
Root	Red root rot	*Poria hypolateritia*	
	Tarry root rot	*Hypoxylon asarcodes*	
	Purple root rot	*Helicobasidium compactum*	
	Charcoal stump rot	*Ustulina zonata*	
	Violet root rot	*Sphaerostilbe repens*	
	Thorny blight	*Aglaospora* sp.	

Table 1.
Commonly prevalent diseases of tea plantation.

the occurrence of blister blight in the year 1868 in North East India [9]. The disease mainly attacks young harvestable tender leaves which are used for the commercial production of tea. Blister blight causes enormous crop loss throughout the major tea-growing countries of Asia including India, Sri Lanka, Indonesia, China, and Japan causing a yield loss of 40% globally [10]. The incidence and severity of blister blight depend on the nature of the tea cultivar, geographical, and environmental conditions of the tea growing areas. The percent yield loss of made tea due to blister blight incidence across the major tea producing countries is represented in **Figure 3b**. Some of the most susceptible tea cultivar prone to blister blight infections are UPASI-1 and UPASI-3 (Assam), UPASI-9 and UPASI-15 (China), and UPASI-17 and TRI-2025 (Cambod), BSS-1, etc. [11]. Here is a comprehensive discussion on the incidence of blister blight disease in different countries, the causal organism, disease cycle, epidemiology, severity, and different approaches employed for the control of blister blight.

1.1 History

Balidon in his book 'Tea in Assam' has indicated the prevalence of blister blight disease on wild indigenous tea in Assam shortly after the beginning of tea cultivation during 1863 [12]. Shortly afterwards Peal in 1868 recognized the existence of blister blight disease of tea and Sir George Watt was the first to report the disease symptoms in Assam in the year 1895. [9, 13]. Later the confirmation of causative pathogen of blister blight as *Exobasidium vexans* was reported by Massee, the Mycologist of Kew Botanical garden in 1898 based on samples sent by Dr. Watt from Upper Assam, India [14]. In 1908, a sudden outbreak of the disease occurred in Darjeeling, West Bengal, India, and

(a)

Climatic condition	Range
Temperature	20-25
Relative humidity	>80%
Sunshine	3.5 h
Rainfall	Moderate
Leaf wetness	15 h

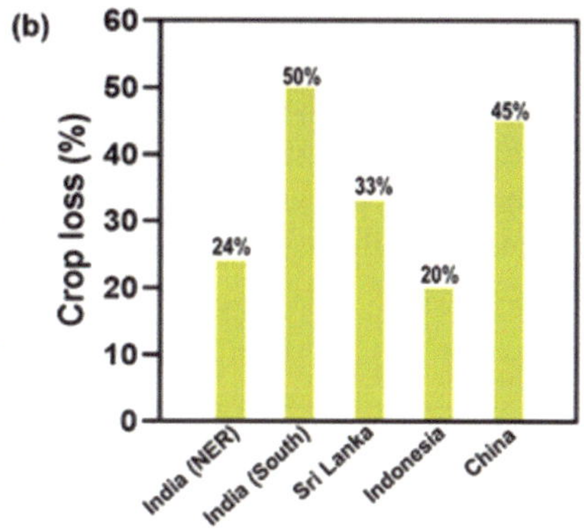

Figure 3.
(a) Climatic condition influencing blister blight disease, (b) blister blight mediated crop loss (%) across major Asian countries [NER: North east region].

was then subsequently reported from Formosa (Taiwan) in 1912, Japan in 1922, and Taiwan in 1938 [15–17]. This was followed by the emergence of the disease in southern India in August 1946, wherein blister blight incidence was first reported in tea estates of Mundakayam and Peermade valley in Kerala. The disease soon spread in the West-Northwest and Southwards direction to Anamallais, other tea estates of Kerala, Nilgiris, Wynaad, and Chikmagalur in Karnataka, due to the effect of South West and North East monsoon winds, thereby affecting the entire tea growing regions of southern India [18]. Later on, the disease was reported from Sri Lanka in 1947 [19], from Sumatra and Java in Indonesia in 1949 [20], Nepal in 1948, East Pakistan in 1951, Thailand in 1953, and South Vietnam and Cambodia in 1959 (CMI, 1970). Hence, blister blight has eventually become a devastating tea disease throughout all the major tea plantations of Asian countries including India, Sri Lanka, Bangladesh, Cambodia, China, Indonesia, Japan, Malaysia, Nepal, Taiwan, Thailand, and Vietnam.

1.2 Blister blight disease symptoms, pathogen, and life cycle

The causal organism of blister blight disease *Exobasidium vexans* is known to be an obligate biotrophic fungus with no alternate host completing its entire life cycle in tea (**Table 2**). The pathogen mainly attacks young, succulent, and tender harvestable leaf and shoot thereby inflicting an enormous effect on the quality and quantity of consumable tea production. The pathogen reproduces through basidiospores which are commonly known to get dispersed by wind. The basidiospores germinated upon lodging on the surface of susceptible tea leaf surfaces under a humid atmosphere with a minimum relative humidity of 80%. The infection is facilitated by the formation of infection peg from appressoria either directly penetrating the cuticle of host tissue or penetration through stomata [21]. The first apparent sign of infection appears in young leaves in the form of pink translucent spots which are considered as the first stage of blister blight infection and are visible after three days of fungal penetration. The spots enlarge along with the leaves and approximately reach a diameter of 3–12.5 mm. In the second stage, the enlarged spots develop into white and velvety convex blister lesions on the lower surface of tea leaf, and on the upper surface, the area with blister lesions becomes sunken resulting in concave depression [22]. The disease progress to its third stage characterized by the curling of the infected tea leaf, browning of blister lesions, and consequently necrosis of the infected leaf tissue. The life cycle of blister blight disease is represented in **Figure 4**. During the off-season, the pathogen is reported to survive on these necrotic leaf parts which facilitate the infection to occur in the subsequent season under favorable climatic conditions. In a study was carried out to detect the survival of *E. vexans* during the off-season the presence of basidiospores in the atmosphere was reported indicating the active state of the pathogen throughout the year. However, the spore concentration being very low (10 spores/m3) the basidiospores failed to sporulate [23]. Under favorable climatic conditions, the pathogen completes its life cycle within 11 days, although it is reported at times extend to 28 days

Kingdom	Fungi
Phylum	Basidiomycota
Class	Exobasidiomycetes
Subclass	Exobasidiomycetidae
Order	Exobasidiales
Family	Exobasidiaceae
Genus	*Exobasidium*
Species	*vexans*

Table 2.
The taxonomic position of E. vexans as described by Massee [14].

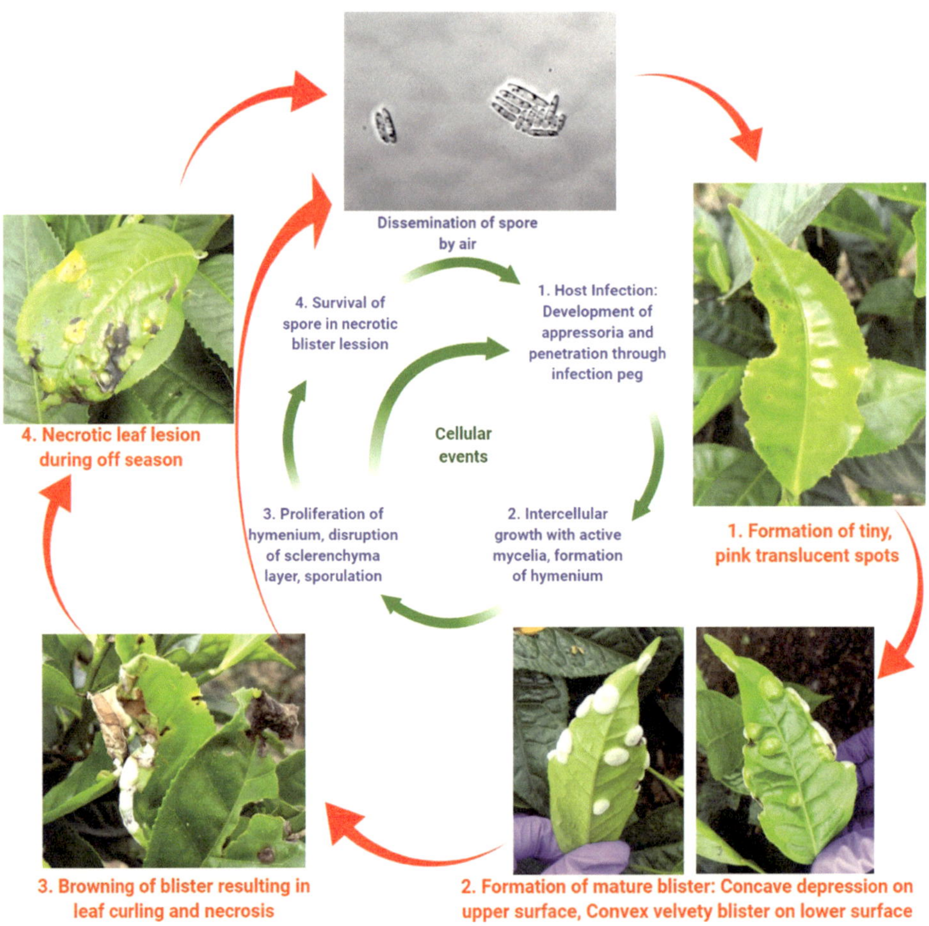

Figure 4.
Disease cycle of blister blight along with life cycle of causal organism Exobasidium vexans.

depending on the prevailing climatic conditions. Owing to the short span of the life cycle, multiple generations of the pathogen are completed within a single cropping season. For the development of blister blight infection sporulation to germination takes place in 4 h to 5 days, germination to penetration takes 4–9 days, penetration to the appearance of visible symptoms takes 3–10 days, development of mature blister and subsequent sporulation takes 11–28 days [11].

A histological study of blister blight disease on tea leaves provided insights into the cellular alteration of host tissue during infection. The study revealed that during the first stage of infection, the enlargement of the translucent spots is a result of hypertrophy as the size of the cells in the mesophyll layer of the infected leaf was substantially higher as compared to the healthy leaf. In the lower epidermis of infected tea leaves with mature blisters, the development of hymenium was prominent in the second stage. This disrupts the lower epidermis completely and gets filled with networks of intercellular hyphae which subsequently develop into basidia that bear basidiospores. However, in tea leaves with blister infections localized in veins, the proliferation of hymenium was apparent in both the lower and upper epidermis. This results in the disruption of the sclerenchyma layer in the vein thereby rupturing xylem and phloem resulting in leaf curling and necrosis of infected leaf in the third stage. The hymenium consisting of bundles of hyphae on maturity forms the clavate to cylindrical basidia (46.98–86.42 μm × 4–5 μm) with normally two and rarely three to four sterigmata [24]. The basidiospores of *E. vexans* are formed at the apex of these sterigmata and two nuclei from the basidium pass into spore via fission. The basidiospores are hyaline and elliptical and measure 7–15.5 X 2.3–4.5 μm when observed under a microscope. It has been reported that 10,000 basidiospores are produced per mm^2 of the blister lesion while the mature blister lesion can produce up to two million basidiospores in 24 hrs [25, 26]. Although the basidiospores are single-celled when immature, three to four septa are reported to form during germination [27]. In a recent study, stages of basidiospore germination were reported under *in vitro* conditions (**Figure 5**). The basidiospores were found to germinate on agar surface 4 h post-incubation followed by germ tube growth from either one or both ends. The spores were initially observed to be aseptate and at a later stage, as many as four transverse septa were formed. At 8 h post-incubation, the formation of hyphae was observed that differentiated into branches to form a complex network of hyphae [28]. So far, the identification of *E. vexans* is being carried out by studying the basidiospore morphology as discussed above, and blister blight disease is identified symptomatically. Molecular based identification remains a major challenge as very few sequences for the ITS region of *E. vexans* have been deposited in NCBI to date. This urges the development of a specific molecular barcode for the identification of *E. vexans*. *E. vexans* being an obligate biotrophic fungus, establishing *in vitro* culture to study the pathogen is another major challenge. In this context, Sundstrom reported that thiamine is a significant supplement in culture media [29]. This was followed by the use of media based on natural substrates for the growth of *E. vexans* under laboratory conditions [30, 31]. Also, vitamin B_5 and calcium pantothenate has been indicated as necessary supplements to maintain *in vitro* culture of *E. vexans* for a period of up to 48 h [32]. Viable *in vitro* culture of *E. vexans* was achieved up to four-five weeks from an optimization study on three different basal media. The carbon source was found as the most significant parameter for czapek dox and v8 juice while tea leaf extract for potato dextrose along with optimal temperature and pH range to be 25–27°C and 7–8 respectively [28].

1.3 Epidemiology, occurrence, and disease severity

The weather condition plays a very important role in the epidemiology and severity of blister blight disease. Low temperature, high humidity, cloudy condition with

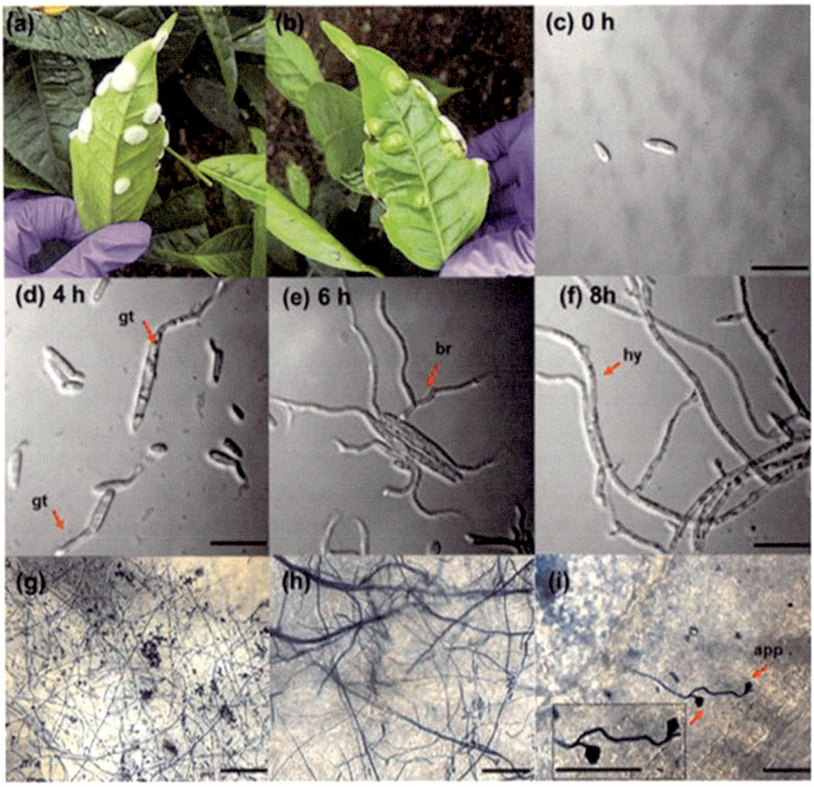

Figure 5.
Representative image of dorsal (a) and ventral (b) surfaces of blister blight infected tea leaf (Ananda tea estate, Lakhimpur, India). Different phases of germination of E. vexans basidiospores on agar: Basidiospores (c); germ tube growth from single end of basidiospores (d); germ tube growth from both ends of basidiospores (e); and hyphal growth of E. vexans with branching (f). Infected leaf part with extensive hyphal growth (g–h) and formation of appressorium (i) (scale: 10 μm, gt-germ tube, br-branching of hyphae, hy-hyphae, app-appresorium) [adapted with permission from Chaliha et al. 2020].

moderate rainfall has been found to play a profound impact on the development of pathogen and disease incidence. As such the incidence of the disease is most favored in monsoon season and facilitated with relative humidity (RH) of more than 80% and availability of water on the leaf surface. In a study carried out by Huysmans (1952), blister blight incidence was recorded with a 5-day average of RH of greater than 83%. On the other hand, Homburg (1953) studied that RH below 80% over 5 days was unfavorable to blister infection. Venkata Ram has reported that the optimum period of leaf wetness to facilitate infection was 11 h and the maximum infection occurred at 13 h [33]. The requirement of moisture content for the germination of basidiospore is reported to be provided by approximately 0.1-inch rain per day while the optimal growth temperature was recorded to be 20–25°C with a maximum tolerance limit of 34°C [29, 34]. Sporulation was found to be inhibited at a temperature greater than 35°C and a temperature of 32°C was reported to be lethal for the basidiospores of *E. vexans* [35]. The incidence of blister blight disease is inversely related to the period of sunshine. Visser et al. in 1961 found a reduction of blister blight disease with an average of 3.5 h of sunshine per day for 5 days at a stretch [34]. Following this report, an exercise of cutting shade trees to medium height for allowing penetration of sunlight to the tea canopy was practiced in Sri Lanka [36]. In a different study, the UV-B (290–320 nm) component of sunlight was found to reduce the sporulation in blister thereby decreasing the number of spores at the end of the disease cycle [37]. Variation in the nature of spores and sporulation behavior was also observed in basidiospores developed under adverse climatic conditions. The basidiospores produced during unfavorable months were found to be thick-walled that failed to germinate. Also, the atmospheric spore count was less during these months. However, in tea plantations close to the ravine that recorded low temperature and high humidity, blister blight disease was noticed during the unfavorable months. Also, tea plants in these areas experienced surface wetness of 15 h which is ideal for blister blight disease incidence [23]. The optimal climatic condition influencing blister blight disease (**Figure 3a**) and percentage crop loss in major Asian countries (**Figure 3b**) is depicted in **Figure 3**.

The spore liberation in the air over blister blight infested tea plantations follows a diurnal rhythm and resembles a nocturnal pattern of spore discharge of other basidiomycete pathogens. The maximum liberation of basidiospores was found to occur between midnight and 4.00 am [38]. The spore deposition on tea plants was found to be directly proportional to the number of spores in the atmosphere. However, the difference in spore deposition in different bushes was observed with higher spore deposition in susceptible hosts [39].

Blister blight disease, being a foliar disease, directly affect the quality and quantity of consumed tea. Severe disease incidence has been recorded after pruning of tea plantations owing to the abundance of young and tender leaf and stem. Also, during infection of tender stem the entire shoot withers and falls along with the curled infected leaf making it unusable for plucking [33]. As such along with enormous yield loss a quality deterioration below 35% disease threshold level is imposed due to blister blight infection [10, 40]. The percentage of crop loss varies with the geographical condition of different countries. In Sri Lanka, Loos reported 50% crop loss in tea plantation without protection, and 33% in plantations protected with copper fungicide [41]. Indonesia reported a loss of ~10 million kg of tea which is 20–25% between 1951 and 1952 [16]. In southern India, during the initial years of blister blight infection, enormous crop losses were observed with an annual loss of about 18 million kg of tea between 1948 and 1952 [42]. North-east India reported, crop loss up to 24% due to blister blight infection, the infection occurring mostly in the hilly region. Darjeeling has been reported with the worst effected tea plantations with blister blight, owing to the favorable climatic conditions. The onset of the disease has been recorded in June with the starting of monsoon and reaches its severity in August till October. In Assam, India blister blight incidence was associated with early rain in February and reaches its severity in the month of March–April [43].

Blister blight infection results in significant degradation of quality in made tea owing to changes in biochemical characteristics [44, 45]. Gulati in 1999 carried out the analysis of biochemical parameters in diseased leaf. In the infected leaves catechin content, total phenols, nitrogen, chlorophyll, amino acids, and polyphenol oxidase activity was recorded in decreasing concentration in comparison to healthy leaf. In orthodox tea processed from infected tea leaves theaflavins, caffeine, catechin polymer thearubgins, and aroma components were significantly found in reduced concentration [46]. Tea shoots with blister infection were also reported with a decrease in catechin content, flavor component 2-phenyl ethanol, and enzyme activity of prephenate dehydrase [47].

1.4 Blister blight disease control

Considering the severity of blister blight disease and its related agronomic and economic losses control of the disease is of utmost necessity. Various control measures have been adopted for the protection of tea plantations against the disease of which the use of therapeutic approaches at the field scale started around 40 years ago [48]. Over the last few years, the various control measures adopted against blister blight disease can be categorized into cultural, chemical, biological, and host tolerance approaches. Different studies have been carried out concerning control measures against *E. vexans* infection. An overall representation of various approaches in controlling blister blight disease is shown in **Table 3**. Details of these approaches and their application at field scale are discussed below.

1.4.1 Cultural practice

E. vexans mainly infect young succulent harvestable leaves which are normally plucked for commercial tea production. The cultural practice of hard plucking also known as fish leaf plucking and early pruning has been employed in tea plantations to reduce the severity of blister blight infection. Eden (1947) reported that hard plucking practice after every two to three months in a year causes no major damage to the tea bushes. However, the long-term hard plucking can result in decreasing crop yield as the bushes get weakened and become susceptible to the attack of mite [57]. Also, tea plantations after hard plucking get delayed and irregular with the growth of new foliage, and as such the tea plantations look worn out [41]. Severely infected tea plants are pruned immediately to control blister blight diseases. Pruning was carried out during hot dry weather to ensure the growth of foliage in the period when the disease presented no danger. However, this resulted in the sunscald damage of the stems that developed into cankers [58]. In line with cultural practices, pruning of shade trees is also rehearsed to allow sunlight to fall on tea bushes as long-term sunlight is reported to inhibit the germination of basidiospores of *E. vexans* [59]. UV-B solar radiation component has been reported to play a significant role in the natural regulation of blister blight disease. The study has shown that removal of UV-B component from sunlight falling on tea bushes resulted in an increasing number of blisters formation while on the other hand, complete sunlight reduced the number of sporulating blisters post 62 h of inoculation. This proves that prolong durations of sunlight can have a negative impact on completing several generations within the same cropping season [37].

Treatment		Disease inhibition	References
Chemical control	Copper oxychloride	85.43%	[49]
	Hexaconazole	78.10%	[49]
	Tridemorph	67.8	[49]
	Propiconazole	78.50%	[49]
	Bitertanol	72.50%	[49]
	Nickel Chloride	84%	[50]
Biological control	*Trichoderma harzianum, Gliocladium virens, Serratia marcescens, Pseudomonas fluorescens, Bacillus subtilis*	40%	[51]
	PGPR	83.94%	[48]
	Ochrobactrum anthropi	73.40%	[52]
	Azoto II-1, *Acinetobacter* sp., Endo-5, Endo-65, and Endo-76	33%	[53]
Systemic acquired resistance	Acibenzolar-S-methyl and salicylic acid	40.80%	[54]
	Calcium chloride ($CaCl_2$)	80%	[55]
	Chitosan	67.70%	[56]

Table 3.
Disease inhibition (%) with various control measures.

1.4.2 Chemical control

The importance of chemical control of blister blight disease and the use of economically feasible chemical therapeutics dates back to 1960 when the disease incidence was recorded in southern India. Protectant fungicides, eradicant fungicides, and systemic fungicides are used as foliar sprays against blister blight disease. Bordeaux mixture and copper oxychloride are the two most commonly used protective fungicide formulations. The acceptance level of the use of copper in tea leaves to control blister blight was set at 150 ppm (Lamb, 1950) as copper-based fungicides also possess collateral damage of phytotoxicity and release of copper residues to environment causes human health hazard, effect soil microflora, and marine population. The formulation of copper oxychloride was able to control blister blight disease at a usage rate of 0.21 Kg metallic copper per hectare. The concentration of copper at 50% wettable powder was used in copper oxychloride formulations [59–61]. Eradicant fungicide nickel chloride hexahydrate was found effective in controlling blister blight disease by antisporulant activity. The reduction in infection was achieved from 84–24% post 3 weeks of treatment and up to 13% after 5 weeks. However, the treatment with nickel chloride was found severely phytotoxic which rejected its use as a potent fungicide [50]. Owing to the phytotoxicity and collateral health hazard from chemical fungicides, organic fungicides were introduced in Sri Lanka, Indonesia, and southern India. However, the disease resistance efficacy was lower in comparison to copper-based fungicides. Also, the high cost related to the processing of organic fungicides discarded its use for blister blight control [25, 60, 62, 63]. Two common brand names for organic fungicides used for blister blight control are Daconil and Difolatan [64].

Conventionally, around 26 rounds of spraying of these fungicides are carried out at 7-days intervals during the disease season to control blister blight incidence. However, since climatic conditions play a significant influence on the severity of blister blight incidence the spraying interval of fungicides differs from region to region. In Indonesia and Sri Lanka an extended period of spraying based on sunshine hours at a specific period of the disease season mediated the control of disease at the economic threshold. On the other hand, control of blister blight disease was not achieved even after a 7-day spraying interval in southern India [65]. Systemic fungicides are often used against plant pathogens owing to its sustained control of plant for example, blister blight could be controlled in southern India by administering pyracarbolid (Sicarol) over 3 weeks. This treatment exhibited strong antisporulant activity reducing the sporulation in mature blisters, while eradicating the latent blister lesions. Also, plant growth was found to be stimulated with the use of pyacarbolid [33]. In a different study with systemic fungicides ergosterol biosynthesis inhibiting (EBI) fungicides tridemorph, bitertanol, hexaconazole, and propiconazole were studied for its effect on physiological parameters of the tea plant and controlling blister blight disease in southern India. EBIs were found with antisporulant activity with a significant reduction in spore size, viability, and inhibited spore germination except for tridemorph treatment. As such, inhibition in spore germination reduced the viability of spore which mediated the reduction in spore load thereby controlling blister blight incidence. The effectiveness of treatment lasted for 7 days, with a reduction of the occurrence of the disease by half relative to untreated plants. Additionally, the EBIs were found with a positive effect on the physiological parameters of the tea plant. The stomatal conductance, total chlorophyll, carotenoids, and photosynthetic rates were found to be induced with EBIs treatment along with an increase in biometric parameters like dry weight and shoot length [49]. In North-East India, 2–3 rounds of systemic fungicides like propiconazole or hexaconazole has been used at 5% EC @ 1:1000 as a foliar spray at 14 days interval to control blister blight infection **(Tea board of India)**.

1.4.3 Biological control

Besides showing appreciable control of blister blight disease with chemical therapeutics, the related phytotoxicity and health hazard have initiated the approach of biological control of blister blight disease. The use of biological control agents like *Trichoderma harzianum*, *Serratia marcescens*, *Gliocladium virens*, *Bacillus subtilis*, and *Pseudomonas fluorescens* have been studied against blister blight disease [66–69]. However, the use of these bioformulations was not found efficient in controlling blister blight disease. Plant growth-promoting rhizobacteria *Pseudomonas* and *Bacillus* were tested for the control of blister blight disease. Foliar application of *Pseudomonas fluorescens* Pf1 bioformulation at 0.5% concentration weekly showed appreciable efficiency against blister blight disease and the lowest mean disease index of 16.06% was achieved which was at par with the chemical fungicide (14.57%). Reduction in disease incidence was achieved for two seasons and the treated tea plants were found with an induced accumulation of defense enzymes peroxidase, polyphenol oxidase, b-1,3-glucanase, chitinase, phenylalanine ammonia-lyase, and phenolics as compared to control. As such, PGPR mediated induced systemic resistance of tea plants against blister blight infestation [48]. Tea phylloplane bacteria have been isolated and screened for their inhibitory action against blister blight disease, of which isolate identified as *Ochrobactrum anthropi* and designated as BMO-111 was found efficient for the biocontrol of *E. vexans*. Foliar application with BMO-111 at 15 days intervals up to 120 days recorded a reduction in disease incidence. Treatment with BMO-111 resulted in 73.4% protection, compared to 64.7% protection with chemical treatments. An inhibitory effect of basidiospore germination and antifungal effectivity was achieved against *E. vexans*. However, the mechanism of action is still unknown and urges further investigation [52]. Although the majority of studies in this context have been carried out in India, a recent study in Indonesia reported the potential use of soil bacteria (Azoto II-1) and three endophytic (Acinetobacter sp., Endo-5, Endo-65, and Endo-76) bacteria against blister blight incidence. The bacterial suspension was to the soil of infected tea plantations at a dose of 2 l ha^{-1}, applied six times at a 1 week interval. The bacterial formulations were observed to control blister blight intensity only up to 2 weeks after treatment. However, the reduction of disease was not significant as compared to control. Also, the treatment was not accompanied by an increase in yield of fresh shoots except for *Acinetobacter* sp. that showed a 17.26% increase in fresh shoot yield [53]. Thus, in the context of biological control of blister blight disease, bioformulation of microbes and its commercialization at the field scale urges further studies for the screening of beneficial microbes that may be applied for efficient management of blister blight disease.

1.4.4 Host resistance against blister blight disease

In the context of blister blight disease, a variety of tea cultivars have been reported that are found to be resistant to *E. vexans* infection. Various studies have been carried out on the morphological characters of these cultivars and the various blister blight defense-related enzymes and pathways have been analyzed to study the nature and basis of resistance. A study was carried out with blister blight resistant clone SA-6 and a susceptible clone TES-34. Both the clones were studied for anatomical differences such as cuticle and epidermal thickness, stomatal length and breadth, palisade tissue, and quantification of epicuticular wax. They have reported that the resistant clone SA-6 showed a higher amount of epicuticular wax, the high thickness of cuticle with greater stomatal frequency when compared to TES-34. Upon pathogen infection, PR proteins are reported to induce systemic acquired resistance in plants. As such, the study also analyzed the difference in the production of PR protein chitinase and have found higher constitutive expression of chitinase in resistant clone [70]. In a recent study with these cultivars, SA6, and TES34, the difference in gene expression against blister blight infection was accessed for chitinase, glucanase, phenylalanine ammonia-lyase, and genes in the flavonoid pathway. The relative intensity of the expression of these genes was found to be higher in the resistant cultivar SA6 in comparison to susceptible cultivar TES34. Also, the expression of these pathogenesis-related genes was found to increase along with each successive stage of blister blight infection [71]. A similar study was conducted recently on the biochemical characterization of resistant tea cultivar AV-2 and susceptible cultivar B-157. Secondary metabolites phenol and proanthocyanidin content were reportedly higher in the resistant clone AV-2. As such, the inferred resistance may be attributed to the antifungal properties of phenol and free radical scavenging activity with the chelation of transition metals by proanthocyanidins. Also, a higher concentration of hydrolyzing enzyme acid phosphatase, peroxidase, catechol oxidase, and superoxide dismutase occur in AV-2 cultivar as compared to B-157 [72].

In line with understanding the basis for resistance of tea cultivar against blister blight disease transcriptome profiling of two cultivars P-1258 (resistant) and T-78 (susceptible) have been carried out to identify defense-related transcripts particularly in a resistant cultivar. cDNA-AFLP mediated the screening of differentially expressed candidate transcripts which mainly showed homology with an acyl-CoA binding protein, zinc finger family protein, ubiquitin, and proline-rich protein that were upregulated after infection. Suppression subtractive hybridization-based transcriptome analysis resulted in a comprehensive study of transcripts induced in resistant cultivar P-1258 after infection. The induced contigs showed similarity to proteins such as ubiquitin family protein, an iron–sulfur cluster scaffold protein, short-chain dehydrogenase, ribulose-1,5-bisphosphate carboxylase/oxygenase small subunit, thioredoxin, pathogenesis-related proteins (chitinase, endo-glucanase, beta-glucosidase, wound-induced protein, protease inhibitor, thaumatin-like protein, cystatin, blight associated protein p12, aspartic proteinase) and proteins with a function in defense signal transduction pathway (serine/threonine-protein kinase, oxo-phytodienoic acid reductase, mitogen-activated protein kinase, leucine-rich repeat transmembrane protein kinase, salicylic acid-binding protein, calcium ion binding or calmodulin-related protein, hydrogen peroxide-induced protein, chitin-inducible gibberellin responsive protein, calreticulin). qRT-PCR based expression analysis of the genes showed greater than two-fold upregulation in P-1258 when compared to T-78 post-infection. Hence, the expression profiling mediated the molecular characterization of resistant tea cultivar involved in developing possible systemic acquired resistance against *E. vexans* infection [73]. To provide further insights into the molecular mechanism of host resistance against blister blight disease, a similar study of transcriptome profiling was carried out with tea cultivars SA6 (resistant) and Kangra-Asha (susceptible) at different stages of blister blight disease. In SA6 cultivar at stage 1 (post 24 h infection), salicylic acid metabolism and secondary metabolite biosynthetic processes were recorded to be enhanced. At stage 2 (7 dpi) hydrogen peroxide metabolic processes, cellular metabolism, and ion transport metabolism-related genes were found to be enriched in resistant variety. As such, hydrogen peroxidase activity suggests efficient scavenging of free radicals enabling restricted penetration/germination of spores inside the host tissue thereby conferring resistance. At stage 3 of infection (14 dpi) increased expression of phenylpropanoid (PAL) and aromatic compound category of gene synthesis probably lead to the synthesis of antimicrobial compounds that might have protected intercellular hymenium development in SA6 cultivar. It was interesting to note that at stage 4 (20 dpi) during necrosis of blister infected part Jasmonic acid-mediated signaling pathway genes are enriched in resistant variety SA6 along with induced monooxygenase and ACC oxidase activity and ethylene production [10]. Here, the induction of both SA and JA signaling pathways again urges to examine the hemibiotrophic existence of the pathogen which is otherwise reported to be biotrophic. In a recent study presence of hypothetical proteins (HPs) was identified and assigned with novel putative defense-related functions in a resistant cultivar of tea SA6 against blister blight disease. The HPs proteins were functionally categorized into LRR, WRKY, NAC, chitinases, and peroxidases. Additionally, different pathways playing a significant role in SA6 resistance against blister blight were annotated based on the KEGG database including plant-pathogen interaction, biosynthesis of secondary metabolites, metabolic pathways, amino sugar, and nucleotide sugar metabolism, and phenylpropanoid biosynthesis which have probably [74].

Plants use various defense mechanisms to shield themselves from infection by pathogens. The cell wall itself acts as an insulation against invading pathogens. Pathogens invading plants breach the cell wall by releasing enzymes and the products get accumulated in the apoplastic region. These are termed elicitors and are capable of activating a complex array of defense signaling called pathogen triggered immunity. These elicitors also mediate the induction of systemic acquired resistance (SAR) in plants. In the last few years, various biotic and abiotic elicitors have been tested and found to mediate SAR in plants against the pathogen. Two chemical elicitors acibenzolar-S-methyl and salicylic acid were tested for their efficiency to induce SAR against blister blight in tea. Plants treated with 0.1% ASM provided 40.8% protection against blister blight. Salicylic acid was used at 250 ppm to achieve significant induction of resistance. Tea plants treated with elicitors were recorded with an induced level of β-1,3-glucanase, phenylalanine ammonia-lyase, and peroxidase activity thereby conferring resistance against blister blight disease [54]. In a similar study treatment of tea plant with abiotic elicitor calcium chloride ($CaCl_2$), found to induce activities of defense enzymes like phenylalanine ammonia lyase (PAL), polyphenol oxidase, peroxidase, and b-1,3-glucanase along with a higher accumulation of total phenolics, thaumatin, cinnamate 4-hydroxylase, flavonoid 30-hydroxylase when compared to control plants [55]. In this context, the use of chitosan as elicitors in tea plants to provide resistance against blister blight disease has been tested and the possible mechanism of resistance has been analyzed. Chitosan solution applied as a foliar spray at 0.01% concentration and 15 days interval reduced blister blight incidence for two seasons. The induced resistance was found to be facilitated by nitric oxide (NO) signaling and the level of total polyphenol content and expression of defense-related enzymes (peroxidase, polyphenol oxidase, phenylalanine ammonia-lyase, and b-1,3 glucanase) was induced [56].

Genetic improvement of tea has been made possible with transgenic technology started from the year 2000. Gene technology and the development in agrobacterium-mediated transformation mediated the incorporation of a foreign gene into crop plants for the development of cultivar with resistance against various diseases. Agrobacterium-mediated transformation was used to develop a disease-resistant variety of tea against blister blight with the introduction of *Solanum tuberosum* class I chitinase gene into tea genome. Plant selectable marker hygromycin phosphotransferase (hpt) gene was used to confer hygromycin resistance. 12 tea plantlets were confirmed with stable integration of transgene that showed resistance against blister blight disease tested in a detached leaf infection assay [75]. In a study carried out by the same group transgenic tea cloned with *Solanum tuberosum* endo-1,3-beta-D-glucanase using Agrobacterium-mediated transformation technique showed significant resistance against blister blight disease. Upregulation of pathogenesis-related (PR) genes like PR3 (chitinase I) gene and PR5 (thaumatin-like protein) gene was recorded in the transgenic tea plantlets which can be attributed to the resistance against blister blight disease [76]. A similar study was carried out for co-transformation of tea genome with LBA4404 pCAMBIA 1301-Chi (carrying potato class I chitinase gene expression cassette) as reported in Singh et al. (2015) and LBA4404 pBI121-Def (carrying mung bean defensin gene expression cassette) together. Although resistance against blister blight was achieved with the co-transformation the resistance was better with *Solanum tuberosum* class I chitinase gene introduced transgene.

2. Conclusion

Widespread research has been carried out so far on understanding the incidence of blister blight infection and on its control measures for the survival of the tea industry. However, to date identification of the pathogen is being carried out symptomatically and morphologically. In the context of molecular identification of *E. vexans* a few sequences for the ITS region are found in the NCBI database which urges to carry out a more detailed study on the development of molecular barcode-based identification. Also, very little is known so far about the genome and transcriptome profile of the pathogen which could be a basis for understanding the molecular mechanism behind the pathogenesis of *E. vexans*. Understanding the molecular basis of pathogenesis of *E. vexans* would likely mediate control measures to be applied more specifically and efficiently. In line with control measures against blister blight, most of the studies reported so far have worked on defense responses and resistance mediated by tea cultivars and hence urges further studies to elucidate the molecular pathogenesis cycle of the pathogen. As such, from the pathogen point of view, there is a significant gap in understanding blister blight disease and a thorough analysis of the pathogen is likely to address the various associated challenges.

Acknowledgements

Author CC would like to acknowledge DST, Govt. of India for her DST INSPIRE Junior Research Fellowship (IF-150964).

Conflict of interest

The authors declare that they have no known potential conflict of interest.

Author details

Chayanika Chaliha[1] and Eeshan Kalita[1,2*]

1 Department of Molecular Biology and Biotechnology, Tezpur University, Napaam, Assam, India

2 Department of Molecular Biology and Biotechnology, Cotton University, Guwahati, Assam, India

*Address all correspondence to: ekalita@tezu.ernet.in

IntechOpen

© 2020 The Author(s). Licensee IntechOpen. This chapter is distributed under the terms of the Creative Commons Attribution License (http://creativecommons.org/licenses/by/3.0), which permits unrestricted use, distribution, and reproduction in any medium, provided the original work is properly cited.

References

[1] Mandal, A. B., Basu, A. K., Roy, B., Sheeja, T. E., & Roy, T. (2004). Genetic Management for Increased Tolerance to Aluminium and Iron Toxicities in Rice—A Review.

[2] Harbowy ME, Balentine DA, Davies AP, Cai Y. Tea chemistry. Critical Reviews in Plant Sciences. 1997;*16*(5):415-480

[3] Sharma RK, Negi MS, Sharma S, Bhardwaj P, Kumar R, Bhattachrya E, et al. AFLP-based genetic diversity assessment of commercially important tea germplasm in India. Biochemical Genetics. 2010;*48*(7-8):549-564

[4] Hazarika LK, Bhuyan M, Hazarika BN. Insect pests of tea and their management. Annual Review of Entomology. 2009;*54*:267-284

[5] Xuefen, C. Z. C. (1989). A world list of pathogens reported on tea plant [J]. *Journal of Tea Science*, *1*.

[6] Chen ZM, Chen XF. In the Diagnosis of Tea Diseases and their Control (Chinese). Shanghai: Shanghai Scientific and Technical Publishers; 1990. p. 275

[7] Lehmann-Danzinger H. Diseases and pests of tea: Overview and possibilities of integrated pest and disease management. Journal of Agriculture in the Tropics and Subtropics. 2000;*101*(1):13-38

[8] Sarkar S, Kabir SE. A field survey of sucking tea Pest and their control measures in few tea Gartdens of Terai region, West Bengal, India. International journal of science and research. 2016;*5*(3):1343-1345

[9] Peal, S.E. (1868). Blister blight. J. Agri, Hort. Soc. India. *1*(126).

[10] Jayaswall K, Mahajan P, Singh G, Parmar R, Seth R, Raina A, et al. Transcriptome analysis reveals candidate genes involved in blister blight defense in tea (Camellia sinensis (L) Kuntze). Scientific Reports. 2016;*6*(1):1-14

[11] Sen S, Rai M, Das D, Chandra S, Acharya K. Blister Blight a Threatened Problem in Tea Industry: A Review. *Journal of King Saud University-Science*; 2020

[12] Balidon, (1877). The tea industry in India. W.H.Allen 8 Co., London

[13] Watt, G. and Mann, H.H. (1903). The pests and blights of the tea plant second edition, Office of the Superintendent, Govt. printing, Calcutta, India. 429

[14] Massee G. Tea blights. Kew Bull. 1898;*1898*:105-112

[15] McRae W. The outbreak of blister blight on tea in the Darjeeling districts in 1908-1909. Agric. J. India. 1910;*5*:126-137

[16] De Weille GA. Blister blight (Exobasidium vexans) in tea and its relationship with environmental conditions. Netherlands Journal of Agricultural Science. 1960;*8*(3):183-210

[17] Liau TL. Blister blight and its control. Taiwan Agric. 1966;*11*:1-5

[18] Subba Rao, M.K. (1946). Blister blight of tea in South India. United Planters' Association of southern India, Coonoor. *4*, 14.

[19] Tubbs FR. A leaf disease of tea new to Ceylon. Tea Quart. 1947;*19*:43-50

[20] Reitsma J, Van Emden JH. De Bladgallenziekte van de Thee in Indonesia. Arch. Voor. *Thee Cult*. 1950;*17*:71-76

[21] Gadd CH, Loos CA. The basidiospores of Exobasidium vexans. Trans. Brit. Mycol. Soc. 1948;*31*:229-233

[22] Punyasiri PN, Abeysinghe ISB, Kumar V. Preformed and induced chemical resistance of tea leaf against Exobasidium vexans infection. Journal of Chemical Ecology. 2005;*31*(6):1315-1324

[23] Ajay D, Balamurugan A, Baby UI. Survival of Exobasidium Vexans, the Incitant of blister blight disease of tea, during offseason. International Journal of Applied. 2009;*4*(2):115-123

[24] Mohktar N, Nagao H. Histological description of Exobasidium vexans infection on tea leaves (Camellia sinensis). Songklanakarin Journal of Science & Technology. 2019;*41*(5)

[25] Loss CA. Tea causative fungus. Tea Quart. 1951;*22*(2):63-72

[26] Huysmans CP. Control of blister blight (Exobasidium vexans) in tea on Sumatra. Bergcultures. 1952;*21*:419-464

[27] Gadd CH, Loos CA. Further observations on the spore growth of Exobasidium vexans. Transactions of the British Mycological Society. 1950;*33*(1-2)

[28] Chaliha C, Kalita E, Verma PK. Optimizing In vitro culture conditions for the biotrophic Fungi Exobasidium vexans through response surface methodology. Indian Journal of Microbiology. 2020;*60*(2):167-174

[29] Sundstrom KR. Studies of the physiology, morphology and serology of Exobasidium. Symb Bot Ups. 1964;*18*(3):1-89

[30] Graafland W. Four species of Exobasidium in pure culture. Acta Botanica Neerlandica. 1953;*1*:516-522

[31] Ezuka A. Artificial culture of two species of Exobasidium, E. vexans Massee and E. japonicum Shirai. Bulletin of Tea Division, *Tokai-Kinki Agricultural Experiment Station*. 1955;*3*:28-53

[32] Nagao H. Effect of aqueous vitamin B on the growth of blister blight pathogen, *Exobasidium vexans*. Songklanakarin Journal of Science & Technology. 2012;**34**(6):601-606

[33] Venkata Ram CS. Calixin, a systemic fungicide effective against blister blight (*Exobasidium vexans*) on tea plants. Pesticides. 1974;*8*:21-25

[34] Visser T, Shanmuganathan N, Sabanayagam JV. The influence of sunshine and rain on tea blister blight, Exobasidium vexans Massee, in Ceylon. Annals of Applied Biology. 1961;*49*(2):306-315

[35] Venkata Ram CS, Chandra Mouli B. Systemic fungicides for integrated blister blight control. UPASI Tea Sci. Pep. Bull. 1976;*33*:70-87

[36] Arulpragasam PV, Addaickan S, Kulatunga SM. Recent developments in the chemical control of blister blight leaf disease of tea - effectiveness of EBI fungicides. *S.L.J.* Tea Sci. 1987;*56*:22-34

[37] Gunasekera TS, Paul ND, Ayres PG. The effects of ultraviolet-B (UV-B: 290-320 nm) radiation on blister blight disease of tea (Camellia sinensis). Plant Pathology. 1997;*46*(2):179-185

[38] Shanmuganathan N, Arulpragasam PV. Epidemiology of tea blister blight (Exoba&idium vexanA) II. The diurnal and seasonal periodicity of spores in the air over a tea estate. *Trans. Brit. Mycol. Soc*. 1966;*49*:219-226

[39] Kerr A, Shanmuganathan N. Epidemiology of tea blister blight (Exoba6ldium ve.Xan.6) 1. Sporulation. *Trans. Brit. Mycol. Soc*. 1966;*49*:139-145

[40] Gulati A, Gulati A, Ravindranath SD, Chakrabarty DN.

Economic yield losses caused by Exobasidium vexans in tea plantations. Indian Phytopathol. 1993;**46**:155-159

[41] Loos, C.A. 1954. Report of the Pathological Division for the Year 1954. Ann. Rep. *Tea Res. Instt. of Ceylon Bull.* No. *36*.

[42] Venkata Ram CS. Blister blight through a decade. UPASI Tea Sci. Pep. Bull. 1961;*20*:38-53

[43] Mann H. Blister blight of tea. Bull Ind. *TeaAsso*. 1906;*3*:1-13

[44] Satyanarayana G, Baruah GCS. Leaf and stem disease of tea in N.E. India with special reference to recent advances in control measures. *J.Pl.* Crops. 1983;*11*(Suppl):27-31

[45] Baby UI, Ravichandran R, Ganesan V, Parthiban R, Sukumar S. Effect of blister blight disease on the biochemical and quality constituents of green leaf and CTC tea. Tropical Agriculture. 1998;*75*(4):452-456

[46] Gulati A, Gulati A, Ravindranath SD, Gupta AK. Variation in chemical composition and quality of tea (Camellia sinensis) with increasing blister blight (Exobasidium vexans) severity. Mycological Research. 1999;**103**:1380-1384

[47] Sharma V, Joshi R, Gulati A. Seasonal clonal variations and effects of stresses on quality chemicals and prephenate dehydratase enzyme activity in tea (Camellia sinensis). European Food Research and Technology. 2011;**232**:307-317

[48] Saravanakumar D, Vijayakumar C, Kumar N, Samiyappan R. PGPR-induced defense responses in the tea plant against blister blight disease. Crop Protection. 2007;*26*(4):556-565

[49] Baby UI, Balasubramanian S, Ajay D, Premkumar R. Effect of ergosterol biosynthesis inhibitors on blister blight disease, the tea plant and quality of made tea. Crop Protection. 2004;*23*(9):795-800

[50] Venkataram CS. Development of spraying technique and usage of fungicide against diseases of tea. Ann. Rep. UPASI Sci. *Dept. Tea. Sect.* 1967;*68*:31-56

[51] Premkumar R, Ajay D, Muraleedharan N. Biological Control of Tea Diseases-a Review. *New Delhi: Research India Publications*; 2009. pp. 223-230

[52] Sowndhararajan K, Marimuthu S, Manian S. Biocontrol potential of phylloplane bacterium O chrobactrum anthropi BMO-111 against blister blight disease of tea. Journal of Applied Microbiology. 2013;*114*(1):209-218

[53] Fauziah F, Setiawati MR, Pranoto E, Susilowati DN, Rachmiati Y. Effect of indigenous microbes on growth and blister blight disease of tea plant. Journal of Plant Protection Research. 2019:529-534

[54] Ajay D, Baby UI. Induction of systemic resistance to Exobasidium vexans in tea through SAR elicitors. Phytoparasitica. 2010;*38*(1):53-60

[55] Chandra S, Chakraborty N, Chakraborty A, Rai R, Bera B, Acharya K. Abiotic elicitor-mediated improvement of innate immunity in Camellia sinensis. Journal of Plant Growth Regulation. 2014;*33*(4):849-859

[56] Chandra S, Chakraborty N, Panda K, Acharya K. Chitosan-induced immunity in Camellia sinensis (L.) O. Kuntze against blister blight disease is mediated by nitric-oxide. Plant Physiology and Biochemistry. 2017;*115*:298-307

[57] De Weille GA. Blister blight Exobasidium vexans in tea and its

relationship with environmental conditions. Netheriand J. Agile. Sti. 1960;*8*(3):183-210

[58] Venkata Ram, C.S. (1974b). Pruning for rejuvenation. Planters1 Chron. 69, 279-282.

[59] Visser T, Shanmuganathan N, Sabanayagam TV. Blister blight control in 1957 with respect to fungicidal formulation application rates and yield. Tea Quart. 1958;**29**:9-20

[60] Jayaraman V, Venkataramani KS. Control of blister blight in tea in southern India. The 1956 field trials. Planters' Chron. 1957;**52**:35-39

[61] Jayaraman V, Venkata Ram CS. Control of blister blight of tea in southernIndia. The 1958 field trials. Ann. Rep. UPASI Sci. Pep. Tea Sect, for. 1959;**1958-59**:28-35

[62] Laoh JP. Fungiclden proeven bij. blister blight (Exobasidium vexans) op. thee. *Arch, Voor*. Thee Cult. 1955;*19*:1-9

[63] Venkata Ram CS. Application of nickel chloride to tea plants (Cameltla &inw6i6) and control of blister blight. Current Science. 1960;**30**:57-58

[64] Venkata Ram CS. Systemic control of Exobasidium vexans on tea with 1, 4-Oxathiin derivatives. Phytopathology. 1969;*59*:125-126

[65] Ram CV, Mouli BC. Interaction of dosage, spray interval and fungicide action in blister blight disease control in tea. Crop Protection. 1983;*2*(1):27-36

[66] Premkumar R. Report of the plant pathology division. Annual Report of UPASI Tea Research Foundation pp. 2001:32-33

[67] Premkumar R. Report of the plant pathology division. Annual Report of UPASI Tea Research Foundation pp. 2002:35-36

[68] Premkumar R. Report of the plant pathology division. Annual Report of UPASI Tea Research Foundation pp. 2003:38-39

[69] Balasubramanian S, Parathiraj S, Haridas P. Effect of vermicompost based Trichoderma (Vermiderma) on the recovery of pruned bushes and on the control of certain diseases in tea (Camellia sinensis (L) O. Kuntze). *J.* Plantation Crops. 2006;*34*(3):524-528

[70] Jeyaramraja PR, Pius PK, Manian S, Meenakshi SN. Certain factors associated with blister blight resistance in Camellia sinensis (L.) O. Kuntze. Physiological and Molecular Plant Pathology. 2005;*67*(6):291-295

[71] Nisha SN, Prabu G, Mandal AKA. Biochemical and molecular studies on the resistance mechanisms in tea [Camellia sinensis (L.) O. Kuntze] against blister blight disease. Physiology and Molecular Biology of Plants. 2018;*24*(5):867-880

[72] Hazra A, Dasgupta N, Sengupta C, Kumar R, Das S. On some biochemical physiognomies of two common Darjeeling tea cultivars in relation to blister blight disease. Archives of Phytopathology and Plant Protection. 2018;*51*(17-18):915-926

[73] Bhorali P, Gohain B, Gupta S, Bharalee R, Bandyopadhyay T, Das SK, et al. Molecular analysis and expression profiling of blister blight defense-related genes in tea. Indian Journal of Genetics and Plant Breeding. 2012;*72*(2):226

[74] Singh G, Singh G, Seth R, Parmar R, Singh P, Singh V, et al. Functional annotation and characterization of hypothetical protein involved in blister blight tolerance in tea (Camellia sinensis (L) O. Kuntze). Journal of Plant Biochemistry and Biotechnology. 2019;*28*(4):447-459

[75] Singh HR, Deka M, Das S. Enhanced resistance to blister blight in transgenic

tea (Camellia sinensis [L.] O. Kuntze) by overexpression of class I chitinase gene from potato (Solanum tuberosum). Functional & Integrative Genomics. 2015;*15*(4):461-480

[76] Singh HR, Hazarika P, Agarwala N, Bhattacharyya N, Bhagawati P, Gohain B, et al. Transgenic tea over-expressing solanum tuberosum endo-1, 3-beta-D-glucanase gene conferred resistance against blister blight disease. Plant Molecular Biology Reporter. 2018;*36*(1):107-122

Chapter 5

Spectroscopy Technology: An Innovative Tool for Diagnosis and Monitoring of Wheat Diseases

Fenfang Lin, Dongyan Zhang, Xin-Gen Zhou and Yu Lei

Abstract

Diseases are among the most important factors limiting worldwide production of wheat. Accurate detection of diseases is the key to develop effective management strategies for control of these diseases. Spectroscopy-based technology can be a non-destructive, quick, efficient tool to accurately detect and monitor the occurrence and development of crop diseases. There has seen an increased interest in the research and application of spectrum technology for the diagnosis and detection of wheat diseases in recent years. This book chapter provides a brief review on research advances in using spectroscopy techniques to detect wheat diseases, with a focus on the diagnosis and detection of Fusarium head blight, powdery mildew, and stripe rust, three important fungal diseases in wheat worldwide. Disease symptoms and traditional disease detection methods are also included. Both literature and our original research data are presented, with the section of conclusion and prospects at the end of this book chapter.

Keywords: wheat, spectroscopy, fusarium head blight, powdery mildew, stripe rust, hyperspectral, remote sensing, vegetation index, fungal disease, symptoms, sensor, *Fusarium graminearum ss*, *Blumeria graminis*, *Puccinia striiformis*

1. Introduction

Wheat (*Triticum aestivum*) is one of the most important food crops in the world [1] and is produced in large areas in many countries, especially in India, China, USA, Europe, and Latin and central America. Wheat ranks third in worldwide food crop production [2]. However, the world's population is expected to increase from 7.8 billion currently to more than 9 billion in 2050 [3]. It has been estimated that global grain production must be increased by more than 60% by 2050 to feed the increased population [4]. Therefore, enhancing wheat production through using high yielding cultivars and improved management practices plays an important role toward this goal. However, numerous diseases, especially powdery mildew (*Blumeria graminis*), brown rust (*Puccinia recondita* f. sp. *tritici*), yellow rust (*Puccinia striiformis*) and Fusarium diseases, pose a challenge to enhance wheat production. Furthermore, recently increased occurrence of extreme weather events pose an additional threat to the global wheat production and food security [5, 6]. Therefore, developing and implementing effective management strategies for wheat diseases is very important to reduce wheat yield and quality loss. However, accurate and quick detection of wheat diseases is the first essential step for effective control of these diseases.

The traditional detection of wheat diseases is achieved by visual observations of farmers that require sufficient knowledge and skills of disease diagnosis or by laboratory tests using destructive sampling method. These methods are tedious, time-consuming, subjective and not suitable for large-scale monitoring. Because of the limitations of these traditional methods, remote sensing technology has gained more and more attentions of researchers since remote sensing is reliable, real-time and precisive [7, 8]. There are three main types of remote sensing systems for plant disease monitoring: 1) visible and infrared spectral systems, 2) fluorescence and thermal systems, and 3) synthetic aperture radar and light detection and ranging equipment systems. Plant disease detection can be based on image or spectroscopic analysis from these sensor systems. A spectroscopy-based method has been used as a wide-ranging approach for the detection of plant diseases and other biotic stresses. Spectroscopy is an interaction of electromagnetic spectrum and matter, which is associated with the type of radiation energy, type of material, nature of interaction, and so on [9]. Examples of molecular and atomic spectroscopy include visible (VIS), infrared (IR), electrical impedance (EI), and fluorescence spectroscopy [10].

The purpose of this chapter is to introduce the readers to the application of spectroscopic techniques on wheat disease detection. This article focuses on three major wheat diseases, Fusarium head blight (FHB, *Fusarium graminearum*), powdery mildew (*Blumeria graminis*) and stripe rust (*Puccinia striiformis* f. sp. *tritici*). The FHB disease, also known as wheat scab, has become one of the most prevalent and damaging diseases in wheat. It can not only affect the normal physiological functions of wheat spikes but also produce deoxynivalenol that is detrimental to both human and animal [11]. Stripe rust and powdery mildew, belonging to foliar diseases, are among the most important diseases and occur in wheat all over the world. Powdery mildew is also one of the most deleterious diseases threatening wheat production worldwide [12]. Stripe rust of wheat is an airborne biotrophic fungal disease and has the capability of fast spread to the new regions and crop cultivars [13, 14]. Compared to leaf and stem rusts, stripe rust is more destructive [15]. Losses in yield caused by these three diseases can reach up to 50% or more depending the year and the country.

2. Principle of spectroscopy diagnosis and monitoring

Crop diseases produce a variety of symptoms and physiological changes in plant tissues, which is the basis for the physical analysis of their spectroscopy characteristics. These external and internal manifestations of diseases can produce an interaction with electromagnetic spectrum or radiation energy. The spectroscopy response of crop diseases can be detected by a specific sensor or sensors system.

All the three wheat diseases have the symptoms of lesions or pustules due to infection (**Table 1**). Single or multiple forms of destruction as illustrated in the table produce different levels of disease severity at the different growth stages, and these symptoms may superimpose or interact with each other. Apart from these visible symptoms, physiological changes appear in plants in the temporal process. For foliar diseases, their infection initially produces lesions on leaves. Without disease control, severer infection leads to the decrease of chlorophyll content and biomass in plants, even impediment of the water metabolism. Distinctive symptoms of each disease can respond in specific spectral regions. The features for a disease of interest can be color space, various fluorescence ratios, reflectance of a waveband or some

Wheat diseases	Plant damages	Symptoms
1. *Fusarium* head blight (*Fusarium graminearum ss*)	White and blighted individual spikelets or the whole spike, pink masses of spores on or in between the spikelets; the green whole crop looks green with white spikes in cases of severe infections; black hard perithecia on infected spikes at the time of harvest; pinkish and shriveled grains.	
2. Powdery mildew (*Blumeria graminis*)	White cottony mycelia with spores covering from fraction to almost the whole leaf area, the presence of cleistothecia as small black bodies (points) interlaced within the light colored at the advanced stages.	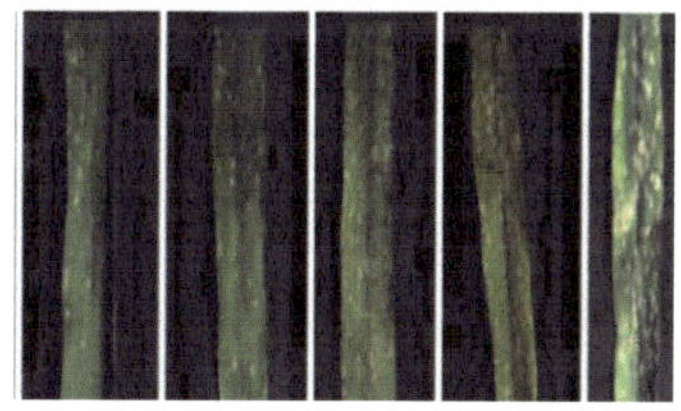
3. Stripe rust (*Puccinia striiformis*)	The formation of uredia in a linear fashion on the leaves and yellow uredinospores on glumes.	

Table 1.
Symptoms of three wheat diseases.

bands, spectral characteristic position variables, or some vegetation indices sensitive to the disease, which depends on the detailed exploration of spectral response mechanism for the disease. Therefore, spectroscopy monitoring can capture spectral information of diseases corresponding to different symptoms and their temporal patterns.

3. Spectroscopic systems available for diagnosing and monitoring wheat diseases

Spectroscopic information for wheat diseases detection can be color space, visible-infrared reflectance spectral and fluorescence parameters, which are derived from digital cameras, multispectral and hyperspectral sensors, and fluorescence devices. Regular digital cameras can obtain Red-Green-Blue (RGB) images with high resolution by ground mobile or Unmanned Aerial Vehicle (UAV) platforms. The leaf or spike color on RGB images is the common indicator of the health status of plants and can be used to measure the chlorophyll content. Multispectral sensors acquire data with more than three bands, including red, green, blue and near-infrared wavelengths. Health plants have high reflectance in the near-infrared region due to multiple scattering resulted from the air-cell interface of leaf inner tissue. Crop diseases can cause reflectivity variation in this region due to water deficit of plants infected with a certain degree of disease. By contrast, hyperspectral sensors have higher spectral resolution and broader wavelength range to detect the more and subtle disease information of crop. This information can indicate morphological and pathological changes of disease-infected plants in more details. Compared to reflectance spectroscopy, chlorophyll fluorescence can provide more direct information about the physiological state of the plants, which is to measure the fluorescence from certain substances after excitation with a beam of light. Studies have found that chlorophyll fluorescence spectrum can sensitively sense plant pathogens at an early stage [16–18].

3.1 Visible digital camera system

A digital camera, covering red, green and blue bands, is the main component of the visible light camera system. It acquires images of the object of interest under the equal angle, illumination and distance. The formats of the images are JPEG, TIF, BMP, and so on. These images can be analyzed by computer vision and texture analysis techniques and plant disease information can be obtained at the image processing stage (**Figure 1**). Image preprocessing is carried out to improve image quality, such as highlighting certain features, noise removal and the enhancement of difference between the object and background. Among them, RGB images are required to convert into color space representation. These color features can be grayscale, Hue Saturation Value (HSV), Hue Saturation Intensity (HSI), Luminosity red-green-blue-yellow (LAB) or YCbCr. LAB consists of luminosity layer (L*), chromaticity-layer (a*), indicating the color fall along the red-green axis, and chromaticity-layer (b*), indicating the color fall along blue-yellow axis. YCbCr is ideal for digital video images processing. In YCbCr format, Y represents luminance information while Cb and Cr are two different color information that stored color information. The originality of chromaticity can be maintained by keeping Cb and Cr constant [19]. Image segmentation is an important step for isolating the leaf or the spike from the background, followed by separating healthy tissue from diseased tissue [20]. Feature extraction involves the selection of optimal color features available for identification of wheat diseases or estimation of wheat disease index. After the features are extracted from the images, classing and staging the disease is the last task in machine vision. The common classifiers used are machine learning algorithms, such as support vector machines (SVM) and artificial neural network (ANN). Many studies have proved the effectiveness of these algorithms, achieving a high classification accuracy for plant diseases recognition under the correct selection of features and classifier parameters [21, 22]. However, success of the disease recognition algorithm depends on many variables that are subject to the judgment of the system designer, such as selection of optimal color space and suitable learning algorithm to use for classification [23]. In recent years, deep learning has been a new trend in machine learning due to its ability to exploit directly raw data without using the handcrafted features. Some studies have used deep learning models for classification between diseased and healthy plant tissues to show its potential to detect plant diseases [24, 25].

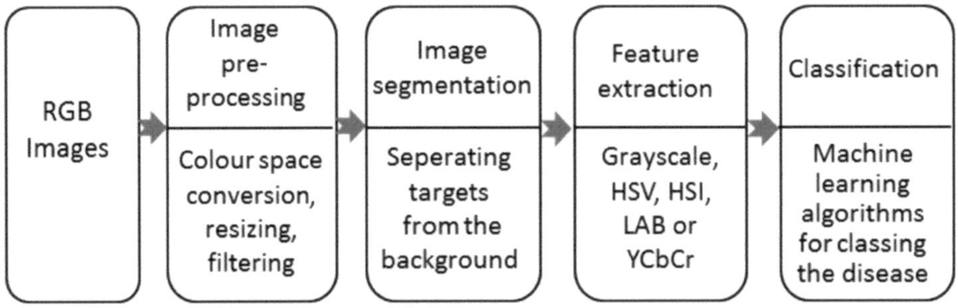

Figure 1.
A workflow for wheat disease detection by RGB images.

3.2 Visible-infrared reflectance spectroscopic system

Multispectral and hyperspectral systems introduced in this chapter are broad-band and narrow-band sensors in the visible and infrared range, respectively. These sensors are mounted on different platforms, such as ground-, aerial-, and satellite-based platforms, which can acquire imaging data with a relatively high signal-to-noise ratio. Given the fact that different symptoms and physiological changes of diseases show specific responses in the reflectance spectrum, monitoring of crop diseases mainly uses the reflected spectral information in data with various spectral resolution and spatial resolution. The general process of crop disease monitoring based on reflectance spectrum involves data preparation, spectral response mechanism analysis, optimal spectral feature selection, detecting model establishment and prevention application (**Figure 2**). Among them, the second and third steps are of importance in spectroscopy-based approaches for crop disease detection. The detailed exploration of spectral response for each disease is the basis of following feature extraction. Each disease can show absorption and reflection peaks in different wavelengths. The narrower the band width, the more obvious and subtler these features. Some sensitive wavelengths for the disease are selected through analysis of the response for construction of a new disease vegetation index. Furthermore, for specific absorption and reflection peak of the disease, a series of spectral characteristic parameters are calculated in these ranges, including position, amplitude, area, ratio and normalized ratio. Some studies show that diseases can cause spectral changes in the regions of the blue edge, green peak, yellow edge, red valley, red edge, and near-infrared. After the construction of specific and optimal indicators for the disease of interest, detecting a specific disease or retrieving the infection severity is conducted by a variety of algorithms or their combinations. For example, some statistical discriminant analysis methods have been adopted and shown good

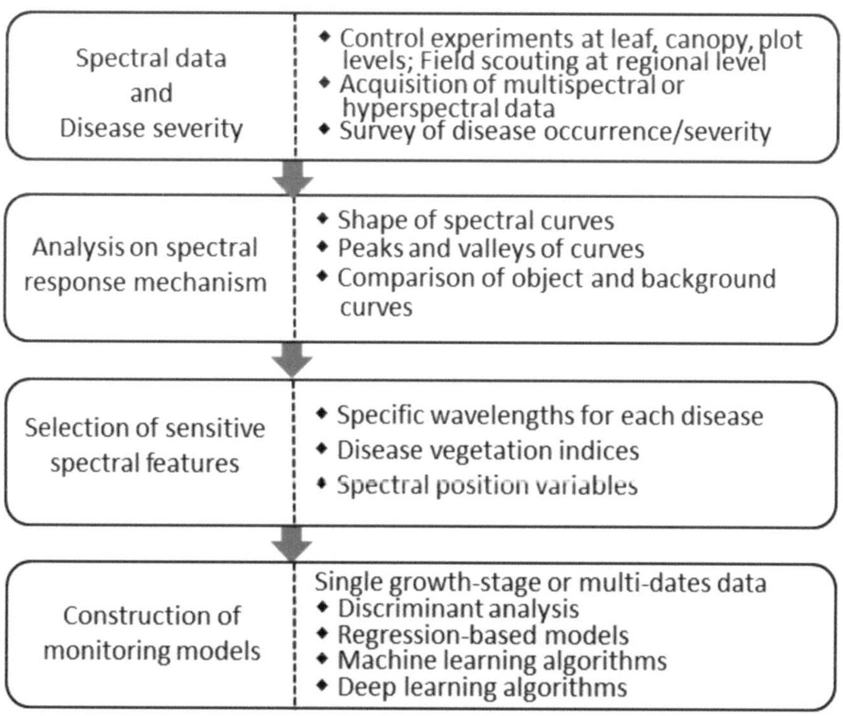

Figure 2.
A workflow for monitoring wheat disease using the VIS-IR spectroscopic technique.

accuracy in detecting one specific disease or differentiating diseases under relatively simple scenarios [26]. And regression-based methods are widely used to determine the infection severity of plant diseases, such as multiple linear regression and partial least squares regression [27, 28]. Compared to these linear statistical methods, many machine learning algorithms, particularly deep learning approaches, have played an important role in modeling for detecting plant diseases under some complicated scenarios [24, 25]. Large amounts of data from multi-phase images can be required when damages caused by diseases are confused with some other factors. This puts forward new and high requirements for modeling.

3.3 Fluorescence spectroscopy system

Chlorophyll fluorescence, as an indicator of plant photosynthesis, can reflect the characteristics of some stresses before infection symptoms are visible, and be used in the monitoring and early warning of various physiological or non-physiological stresses [10]. Active-induced chlorophyll fluorescence techniques have been widely studied due to their strong intensity and abundant information. These fluorescence systems can obtain chlorophyll fluorescence spectrum, chlorophyll fluorescence kinetic parameters, chlorophyll fluorescence image, and life span for detecting plant stresses, such as the stresses of water, high temperature, nutrition, diseases, insect pests, and so on. Fluorescence spectroscopy devices include fiber-optic fluorescence spectrometer, imaging multispectral and hyperspectral sensors. Various ratios of fluorescence amplitude at fluorescence peaks are determined from the continuous fluorescence spectra for achieving pre-symptomatic detection for some pathogens. For example, three fluorescence ratios (F451/F522, F451/F687, F451/F736) have been used to detect early powdery mildew infection in susceptible and resistant cultivars in wheat [29]. The fluorescence spectra over 370–800 nm is also useful for pre-symptomatic detection of wheat leaf rust [30]. In addition, many types of fluorescence kinetic parameters are constructed and used based on the saturation pulse fluorescence analysis, such as the maximum quantum efficiency of photosystem II (PSII) primary photochemistry (Fv/Fm), the maximum efficiency of PSII photochemistry in light adapted material (Fv'/Fm'), non-photochemical quenching (NPQ), and the effective quantum yield of photosystem II (ΦPSII) [31, 32]. These parameters can be obtained by the continuous excitation (FRR) chlorophyll fluorescence detection system and the saturation pulse modulation (PAM) fluorescence dynamics detection system.

4. Detection and diagnosis of wheat FHB

Fusarium head blight (FHB) is a major and devastating disease of wheat worldwide. The pathogen invades wheat ears and causes damage in the kernels in the form of atrophy, weight reduction, and discoloration, resulting in significant yield and quality loss. The fungus also produces mycotoxin deoxynivalenol (DON) that could destroy healthy cells by inhibiting protein synthesis. Food and feed contaminated with DON pose a serious threat to human and animal health [33, 34]. Management practices for control of FHB and DON contamination is critical to the profitable and sustainable production of wheat. Detection and diagnosis of the development of this disease is the first but essential step toward control of it.

4.1 Hyperspectral data acquisition and assessment of disease severity

The hyperspectral images of wheat ears were acquired by the SOC710E imaging spectrometer (Surface Optics Corporation, San Diego, USA) in the field. This imager has a spectral range of 400–1000 nm and the spectral resolution of 2.3 nm. Wheat ears in the field were cut and put on a black light-absorbing cloth. Then, the data were collected by the hyperspectral imager from11:00 AM to 2:00 PM under fine weather conditions (**Figure 3**). The images obtained contained the front side (defined as side A) and the reverse side (defined as side B) of wheat heads for making sure the comprehensiveness of data acquisition.

After hyperspectral data collection, disease severity level (DSL) of wheat FHB was calculated according to the GBT 15796–2011 Rules for Monitoring and Forecast of the Wheat FHB.

$$DSL = \frac{W_d}{W_h} \tag{1}$$

where W_d is the number of pixels within the diseased area of the wheat ear; and W_h is the number of pixels in the whole ear after removing the wheat awn.

The lesion area in wheat ear was extracted by the binarization and morphological processing (**Figure 4**). The original images were RGB images combined by the three central wavelengths of red (660 nm), green (560 nm) and blue (480 nm) from hyperspectral images of wheat ears. The RGB image was then converted to another color space with three components of Y, Db and Dr. Finally, a threshold segmentation method was used to extract and calculate the diseased area of wheat ears.

4.2 Spectral response of wheat ears under different disease severities

In the process of using hyperspectral data to accurately identify diseases, the study of spectral signatures under different disease severities is the basis for screening and identifying sensitive bands of diseases. **Figure 5** shows the spectral response of wheat ears with different infection levels. Generally, in the range of the 550–720 nm band, the spectral reflectance of healthy ears is lower than that of infected samples, with an

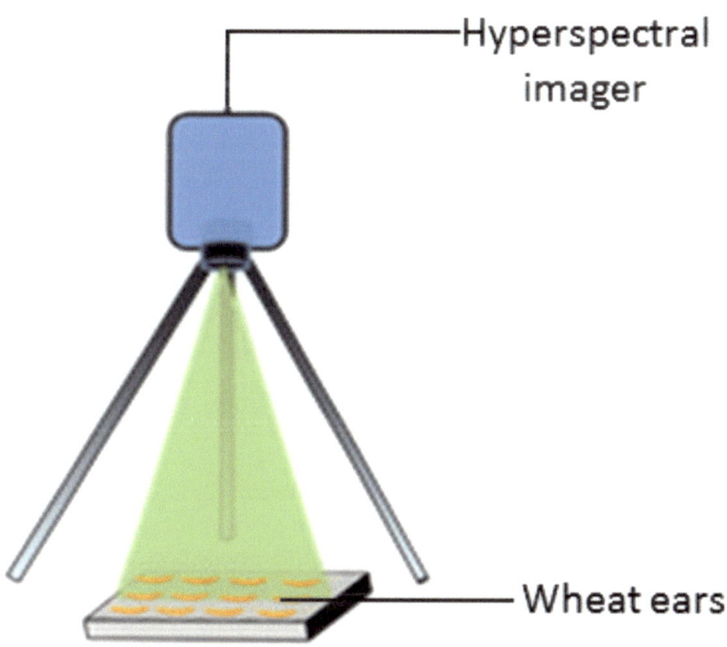

Figure 3.
Acquisition of hyperspectral images for wheat ears.

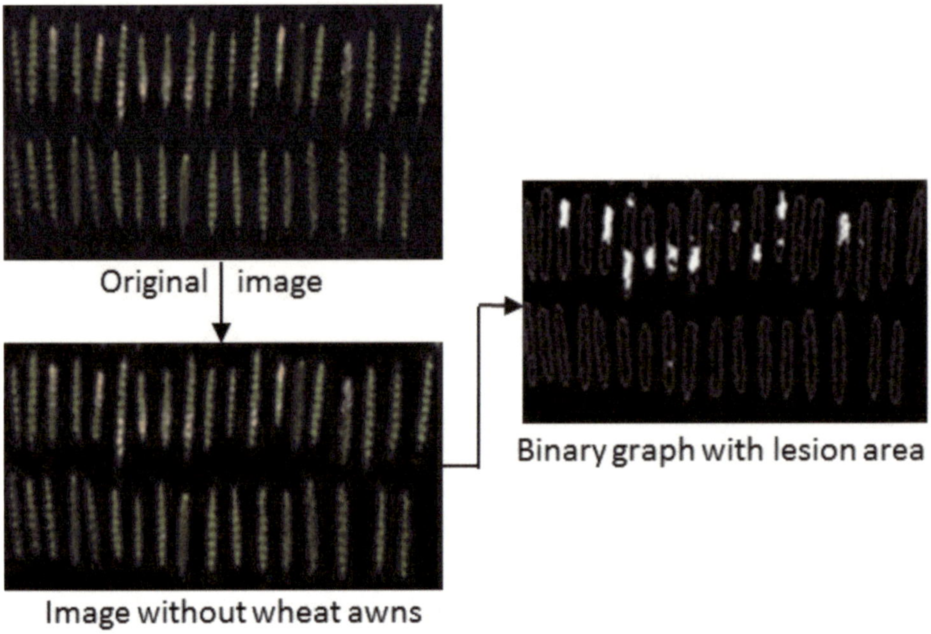

Figure 4.
Extraction of diseased areas from wheat head: (A) wheat head image, (B) wheat head image with wheat awns removed, and (C) image of diseased areas extracted.

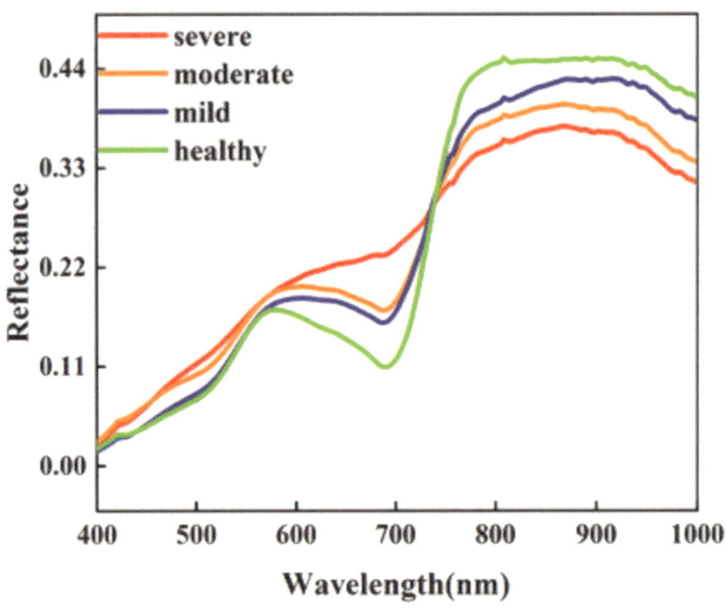

Figure 5.
Spectral reflectance curves of wheat ears with different disease severities.

obvious green peak and red valley. Accordingly, these two spectral features disappear in the severely infected ears. Conversely, in the range of the 721–1000 nm band, the more severe the infected wheat ear the lower its reflectivity. The difference in responses of wheat ears with different severities in the bands of 550–720 nm and 721–1000 nm may be related to pigment and water content in mesophyll tissue [35]. Furthermore, with the increase of disease severity, the red edge position has a trend towards shorter wavelengths. The above obvious spectral signature differences provide an important optical basis for analyzing and constructing the relationship between the spectral index and FHB severity in this work.

4.3 Prediction of FHB severity based on novel spectral disease index

4.3.1 Wheat material

The tested wheat cultivar was Xinong 979, moderately susceptible to the FHB disease. In this work, 149 and 229 wheat ears were collected at the late flowering (3 May 2018) and early filling (9 May 2018) stages for a total of 378 samples.

4.3.2 Characteristic band selection by random Forest

The algorithm of random forest (RF) was used to select the characteristic wavelengths that are sensitive to wheat FHB. The weight coefficients of all wavelengths were calculated in the spectral range of 400–1000 nm. To reduce the redundant information and maximize the effective spectral information, this work selected the wavelengths corresponding to the positive highest and negative lowest weight coefficients as the characteristic wavebands. As shown in **Figure 6**, the characteristic wavelengths were 570 nm and 678 nm at the late flowering stage, 565 nm and 661 nm at the early filling stage, and 560 nm and 663 nm at the combined stage.

4.3.3 Construction of proposed new spectral disease index for identifying wheat FHB

Previous studies [36, 37] have shown that the disease spectral index in the form of the normalized wavelength difference is very sensitive to spectral changes caused by powdery mildew, stripe rust, and aphids. Therefore, this work used the normalized wavelength difference in combination with characteristic wavelengths to construct the exclusive fusarium disease index (FDI) for each period. The formula is as follows.

$$FDI = \frac{R_{\lambda 1} - R_{\lambda 2}}{R_{\lambda 1} + R_{\lambda 2}} \tag{2}$$

where $R_{\lambda 1}$ represents the reflectance at the λ1 wavelength, and $R_{\lambda 2}$ represents the reflectance at the λ2 wavelength.

Here, with FDI as the independent variable and disease severity level (DSL) as the dependent variable, the relationship between FDI and DSL in different stages was

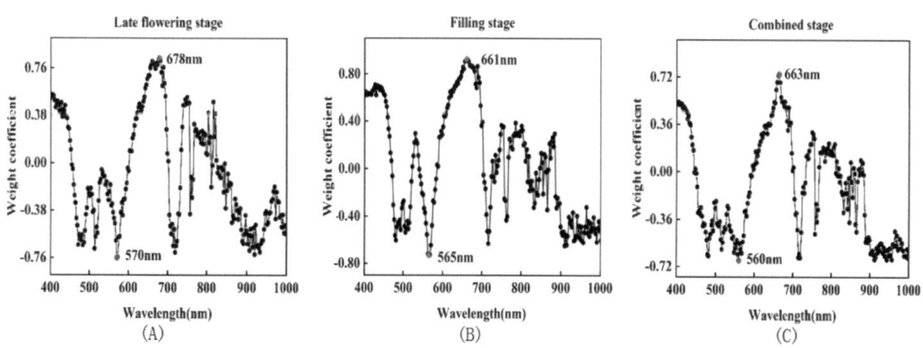

Figure 6.
Weight coefficients calculated by RF at the late flowering (A), filling (B), and combined (C) stages.

evaluated by linear regression analysis (**Figure 7**). FDI made an accurate prediction of the DSL of wheat ears at the late flowering stage, filling stage, and combined stage (R^2 was greater than 0.90, RMSE was less than 0.08). At each stage, the R^2 and RMSE values of the training and test datasets were close, indicating that the model had a strong generalization ability. In addition, the prediction accuracy of FDI was the highest at the filling stage, followed by the late flowering stage and the combined stage.

In this work, the regression model obtained at the combined stage was applied to the test datasets at the late flowering and filling stages, respectively (**Figure 8**).

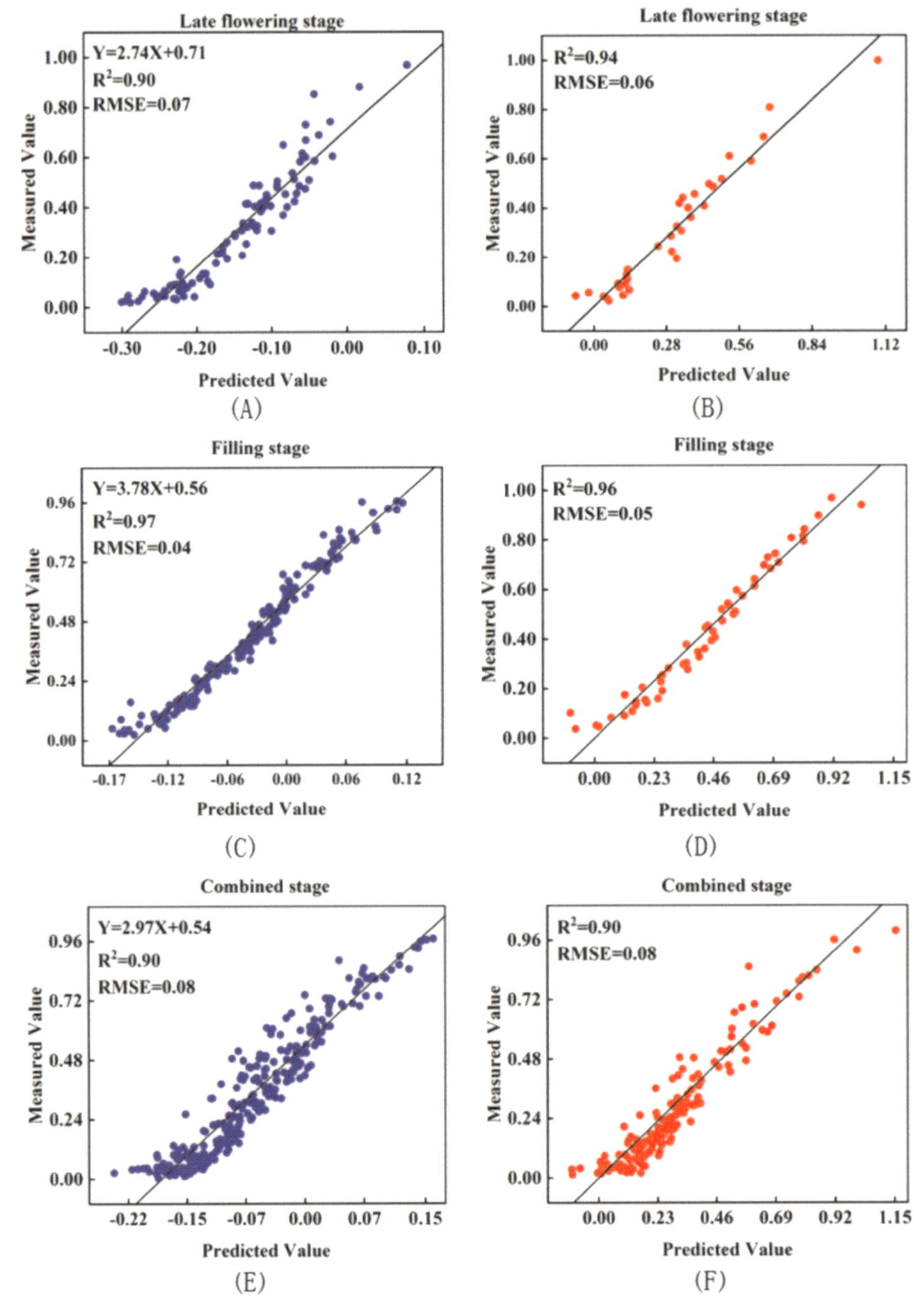

Figure 7.
Regression models of the training datasets at the late flowering (A), filling (C), and combined (E) stages; and verification results of test datasets at the late flowering (B), filling (D), and combined (F) stages.

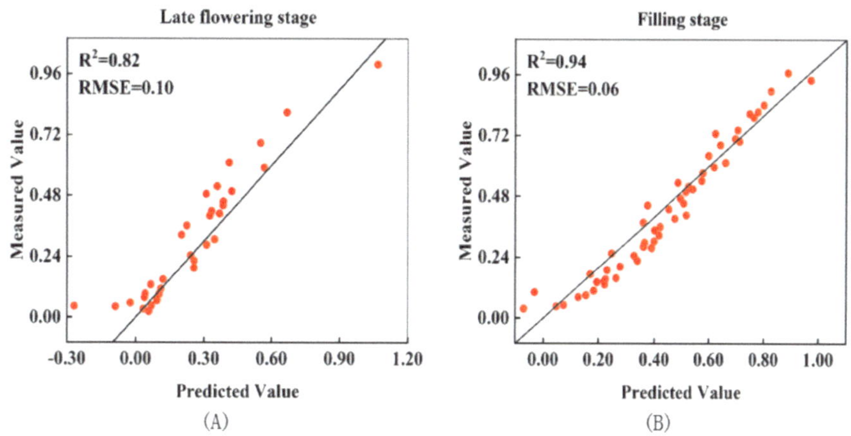

Figure 8.
Results of applying the regression model of the combined stage to test data of the late flowering stage (A), and the filling stage (B).

It was clear that the prediction accuracy of the model was close to that of the model established at the filling stage (R^2 = 0.94 and 0.96, respectively), but lower than that of the model established at the late flowering stage (R^2 = 0.82 and 0.94, respectively).

4.3.4 Comparison of FDI and commonly spectral index

Sixteen commonly used spectral indices were selected and compared with fusarium disease index (FDI) proposed in this work to evaluate the ability of FDI to identify and distinguish infected ears [38]. At the late flowering and combined stages, only FDI had an R^2 more than 0.9 in the training and test datasets. At the filling stage, only FDI and nitrogen reflectance index (NRI) had an R^2 above 0.9 in both the training and test datasets. The characteristic wavelengths of FDI (661 nm, 565 nm) and NRI (670 nm, 570 nm) were also close. The prediction accuracy of FDI to the disease severity was higher than that of other spectral indices at each stage, especially at the late flowering stage. Furthermore, transformed vegetation index (TVI), green index (GI), plant senescence reflectance index (PSRI), normalized difference vegetation index (NDVI), optimized soil-adjusted vegetation index (OSAVI), Lichtenthaler's indices (Lic1) performed relatively well at the late flowering and filling stages, but not at the combined stage. This may be attributed to the change of characteristic wavelengths at different growth stages.

4.4 Integrating spectral and image data to detect wheat FHB

4.4.1 Wheat material

Samples of 1,680 wheat ears with different levels of FHB severity, including the front side (defined as side A) and the reverse side (defined as side B) of wheat heads, were randomly collected from the healthy and infected samples in the research plots at the middle-to-late flowering (1 May 2019) and maturity (11 May 2019) stages.

4.4.2 Extraction process of spectral and image features

Spectral and image features of FHB were extracted from hyperspectral data of wheat ears (**Figure 9**). Spectral reflectance of each wheat ear was converted from the

(A)

(B)

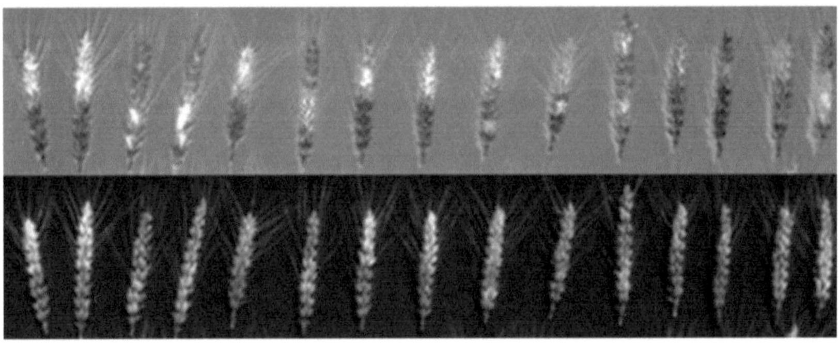

(C)

Figure 9.
Different feature images of wheat ears with FHB: (A) RGB image; (B) YDbDr image; and (C) images of the first second principal components (PC1-PC2).

intensity data of each sample using the ENVI 5.3 software (ITT Visual Information Solutions, Boulder, Utah, USA). Hyperspectral data has a large number of wavebands with information redundancy. Thus, principal component analysis (PCA) was used to reduce the high dimensional spectral data to a few principal components which can retain as much information as possible. The results showed that the first six principal components (PC1-PC6) explained over 98% of the variance and contained most of the information in the hyperspectral image.

Image information used for identifying wheat FHB were texture and color features. The work adopted gray level co-occurrence matrix (GLCM) and dual-tree complex wavelet transform (DTCWT) to obtain texture information of wheat FHB. The texture parameters based on GLCM contained energy, contrast, correlation, entropy and homogeneity calculated from four orientation values, respectively [39]. DTCWT can generate a quadtree structure, including two trees for the image rows and two trees for columns [40–42]. Color information was acquired from RGB images by computer vision technology. The RGB image was composed of red (660 nm), green (560 nm) and blue (480 nm) from hyperspectral images of wheat ears, and transferred to the YDbDr color space. Then, three moments were extracted from RGB and YDbDr space.

Spectral and image features obtained by above-mentioned methods were not all sensitive to wheat FHB. The algorithm of gradient boosting decision tree (GBDT) was used to select optimal features. GBDT is a common nonlinear model by scoring various features and removing the lowest score from features by the sequential backward elimination [43]. Finally, the number of optimum feature subsets was

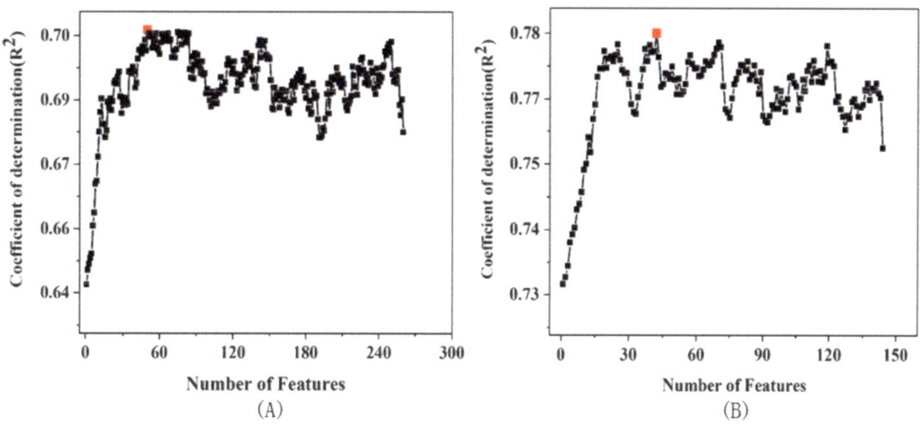

Figure 10.
The relationship between the number of features and R^2: (A) spectral characteristics in the side A of wheat ears, (B) image characteristics in the side A of wheat ears.

determined by the coefficient of determination (R^2) (**Figure 10**). The value of coefficient of determination (R^2) rose when the spectral features with the lowest scores were deleted (**Figure 10(A)**). Obviously, the model performance was improved due to redundant features being deleted. However, R^2 greatly decreased with the deletion of the useful features when the number of important features was reduced to one. For image features (**Figure 10(B)**), there was a similar trend to that of spectral features. Thus, the first 50 spectral and 40 image characteristics were selected from these features, respectively.

4.4.3 Identification of wheat FHB severity

The algorithm of random forest was used to establish the recognition model of wheat FHB with different patterns of feature combination (**Table 2**). The R^2 values of the models based on the spectral, image and fusion features increased successively. The performance of the model by spectral information was less effective than that of the model by image information for both sides of wheat ears. Even after reducing dimensions, the model with image features still had higher R^2 value and lower root mean square error (RMSE). However, the model constructed by the fusion features with spectral and image information provided the best performance, the R^2 values of which were 0.82 and 0.78 for data of side A and side B in wheat ears, respectively. Furthermore, the model by all data of side A and side B had R^2 of 0.89 and RMSE of 6.52 when using fusion features. It was indicated that the fusion

Features	**Side A of wheat heads**		**Side B of wheat heads**	
	R^2	RMSE	R^2	RMSE
Image	0.75	9.08	0.70	9.81
Spectral	0.68	10.25	0.63	10.98
Image- reduction	0.78	8.47	0.73	9.80
Spectral- reduction	0.70	10.30	0.64	11.70
Combined	0.82	7.98	0.78	8.46

Table 2.
The prediction accuracy of random forest models with different features.

of spectral and image information can improve the performance of the identification model of FHB [44].

5. Hyperspectral monitoring of early powdery mildew in wheat

The detection and differentiation of powdery mildew at early stages not only allow timely fungicide treatment to reduce yield loss but also reduce fungicide usage to delay the build-up of fungicide resistance [45, 46]. Some studies have been conducted to evaluate the potential of using hyperspectral remote sensing to detect and quantify powdery mildew [47, 48]. The objective of our study was to construct a spectral disease index for early detection of powdery mildew in wheat.

5.1 Experiment and hyperspectral data acquisition

The experiment was conducted in a field located at the Beijing Academy of Agriculture and Forestry Science, China (39°56'N, 116°16'E). The cultivar of winter wheat used was 'jingshuang 16', highly susceptible to powdery mildew. This was a popular cultivar, and widely grown in China. Hyperspectral data were acquired from wheat leaves. Leaf samples were collected in the field at the filling stage and transported to a nearby laboratory for imaging spectrometer data measurement.

Imaging spectrometer data of leaves were obtained by the customized visible and near-infrared hyperspectral imaging system. The components of the system were shown in **Figure 11**. The spectral range of the system was between 400–1000 nm and its spectral resolution was 2.8 nm. The samples passed its view slot by the electric moving stage when the CCD camera was fixed over the stage. In this process, there were appropriate intensity of illumination and exposure time of the camera. Then, the images were acquired line-by-line.

Due to image noises in the wavebands with weak spectral response, spectral calibration was performed for the dark current correction to eliminate parts of the data noises. In addition, the spectral curve was normalized by band reflectance divided by the mean reflectance of the spectrum to suppress illumination changes among different measurements [49].

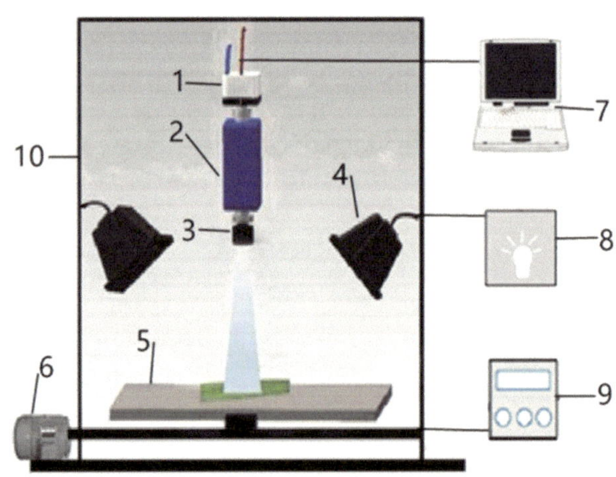

Figure 11.
The hyperspectral imaging system consisted of: (1) CCD camera, (2) imaging spectrograph, (3) lens, (4) light source, (5) sample stage, (6) electric moving stage, (7) computer, (8) light source controller, (9) moving stage controller, and (10) dark room.

5.2 Determination of disease index

Disease index (DI) was used to describe the levels of powdery mildew severity. The DI value of each leaf sample was determined by two processes. Firstly, the severity of wheat powdery mildew was estimated by the percentage of lesion area to whole leaf area and visual estimate of the percentage of infected leaf surface area on the leaf blade. Secondly, the estimated severity was divided into four categories to minimize human error for obtaining the disease index (DI), referring to Chinese Standard (NY/T 613–2002): 0–3% (no disease), 3%–10% (DI = 1), 10.1%–20% (DI = 2), and 20.1%–30% (DI = 3).

The calculation process of disease severity was shown in **Figure 12**, mainly including leaf area extraction and lesion region segmentation. By comparing the pixel values of a leaf blade in RGB color space, it was found that the pixel value of R component was the largest in the lesion area, while the pixel value of G component was the largest in the normal leaf area. Therefore, the ultra-red color feature 2R-G-B was selected for segmenting the normal leaf from other background (disease spots or insect injury). Images at 680 nm, 550 nm and 450 nm were synthesized pseudo color images. The automatic threshold of a component and b component in lab color space was set to extract the leaf area, and then the ultra-red color feature 2R-G-B was used to segment the disease spots on the leaves.

5.3 Construction of powdery mildew index (PMI) by the RELIEF-F algorithm

Powdery mildew index (PMI) consisted of a relevant single wavelength and a normalized wavelength difference. It was shown as the following formula.

$$PMI = \frac{R_1 - R_2}{R_1 + R_2} \pm 0.5R_3 \tag{3}$$

Where, R_1 and R_2 are reflectance in normalized wavelengths, and R_3 is reflectance in the most relevant single wavelength.

The most relevant single wavelength and normalized wavelengths were obtained by the RELIEF-F algorithm. The RELIEF-F algorithm was designed to measure how well attributes distinguished between instances within the close proximity of each

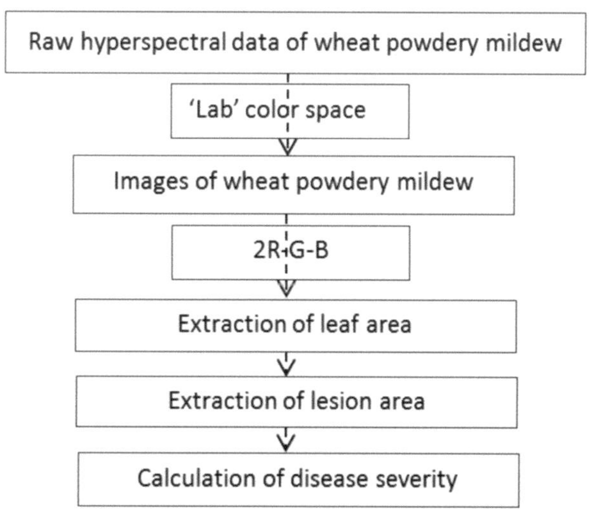

Figure 12.
The calculation process of disease severity.

other [50, 51]. Therefore, the algorithm was useful in finding some wavelengths for specific diseases. According to the study by Huang et al. [36], the most sensitive single wavelength was among wavelengths of the highest weighted (20%). Two normalized wavelengths were wavelengths from the best and worst weighted wavelengths (10%), respectively. And the distance between the two wavelengths was less than 50 nm.

5.4 Detection of early wheat powdery mildew by PMI

Figure 13 shows curves of normalized spectra and first-derivative spectra. The shapes of spectral curves of normal and diseased leaves for normalized spectra and derivative spectra were similar, but several differences existed in specific wavelength ranges among the three disease severity categories. At 450–700 nm, the reflectance value of normal leaves in red edge region had minimal value, followed by very slight and slight leaves had the highest value. At 750–1000 nm, normal leaves had highest reflectance value, followed by very slight leaves and slight leaves had lowest reflectance value. The more serious the disease was, the higher the

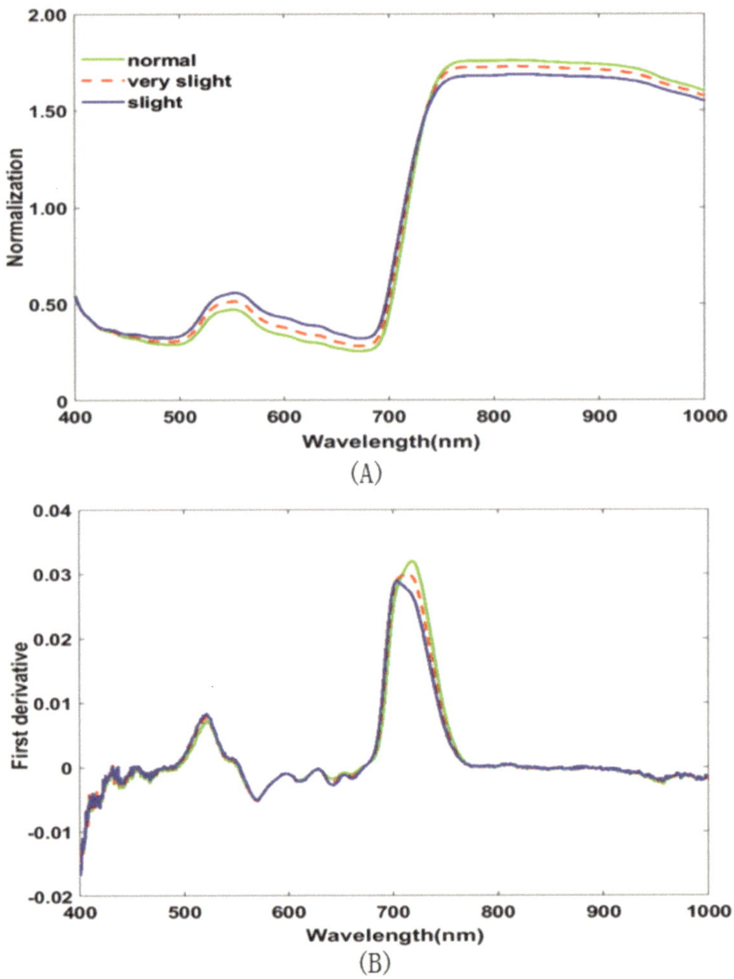

Figure 13.
Curves of normalized spectra and first derivative spectra. (A) Normalized reflectance curves of normal, very slightly-damaged (3% < lesion portion <15%) and slightly-damaged leaves (15% lesion portion 30%); and (B) first derivative spectral curves of normal and diseased leaves.

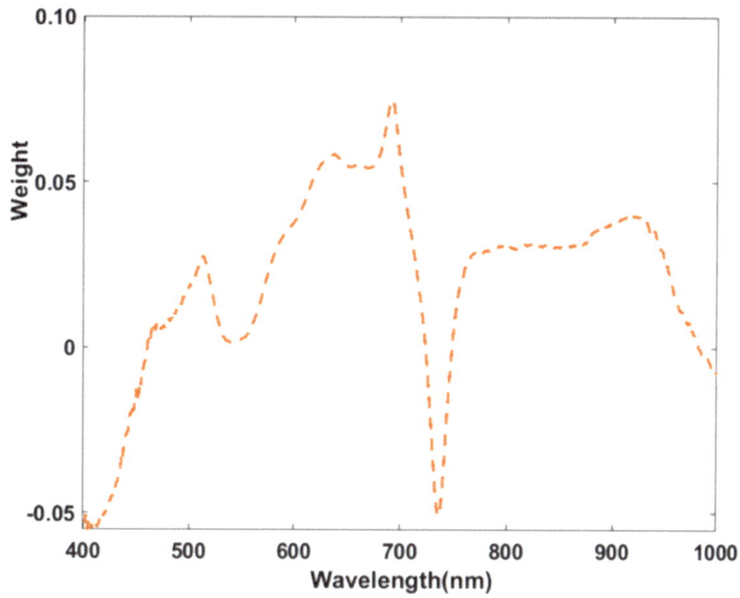

Figure 14.
Single wavelength weights for regression analysis of PMI with disease severity.

reflectance of leaves in the visible region was, while the opposite was true in the near infrared region. These results were consistent with previous studies in metaphase of powdery mildew [12, 52]. However, differences of the first derivative spectra between normal and diseased leaves were significant in the red edge region from 690 to 740 nm, but not obvious in the green edge region from 510 to 530 nm.

The curve of single wavelength weights is shown in **Figure 14**. Single wavelengths around 690, 820, and 910 nm had high relevance to diseased winter wheat

Ranking	Indices	Coefficient of correlation	Formula	Citation
1	**PMI**	**0.90**	**(R686 - R443)/(R686 + R443)-0.5R913**	
2	NBNDVI	-0.86	(R850 - R680)/(R850 + R680)	[48]
3	ARI	0.85	(R570 - R670)/(R570 + R670)	[49]
4	PSRI	0.84	0.5[120(R750-R550) - 200(R670 - R550)]	[50]
5	TVI	-0.77	(R531 - R570)/(R531 + R570)	[51]
6	MCARI	0.76	(R550 - R531)/(R550 + R531)	[51]
7	PRI	-0.76	(|(a670 + R670 + b)|/(a2 + 1)1/2) - (R700/R670) a = (R700 - R550)/150, b = R550 - (a - 550)	[52]
8	CARI	0.72	3[(R700 - R670) - 0.2(R700 - R550)(R700/R670)]	[53]
9	TCARI	0.57	[(R701 - R671) - 0.2(R701 - R549)]/(R701/R671)	[54]
10	RVSI	0.56	[(R712 + R752)/2] - R732	[55]
11	PhRI	0.22	(R680 - R500)/R750	[56]
12	NRI	-0.17	$(R550)^{-1} - (R700)^{-1}$	[57]

R686 stands for reflectance at the band of 686 nm; and others stand for reflectance at their corresponding bands.

Table 3.
Summary of correlation analysis between vegetation indices and DI [58–62] .

leaves, and the normalized reflectance differences were around 445 and 680 nm. All possible wavelength combinations were calculated for the specific spectral index by the RELIEF-F algorithm. Finally, PMI for estimating DI was adopted based on reflectance at 913 nm and normalized reflectance difference between 443 and 686 nm. Statistical correlation analysis was carried out between PMI and DI. It was found that DI had a significant positive correlation to the PMI (R^2 = 0.798, n = 70). This indicates that the PMI had the potential for monitoring the early development of powdery mildew.

To compare the estimating ability of disease index between PMI and other hyperspectral vegetation indices, **Table 3** summarizes the results of correlation analysis between each of the 12 spectral features and DI. Of them, four indices had the coefficient of correlation more than 0.8, which were PMI, NBNDVI, ARI and PSRI. However, PMI had the highest value, indicating that PMI had powerful sensitivity to leaf powdery mildew.

6. Detection and diagnosis of wheat stripe rust

Wheat stripe rust (*P. striiformis* f. sp. *Tritici*) is one of the most destructive diseases of wheat worldwide, with characteristics of high prevalence, explosive occurrence, and severe damage at large areas. This disease is widespread in eastern and central Asia, western Europe and the Pacific coast of North America as well as Oceania, northern and eastern Africa, and south America. The infection process of the disease can be divided into the phases of contact, invasion, incubation, and occurrence. The spread of the spores in the subject cannot be observed by the naked eye during the incubation period. However, wheat stripe rust can easily become epidemic over a large area under suitable conditions once it enters the occurrence phase. Therefore, it is important to have a method that could rapidly and accurately detect stripe rust during the incubation period. Once it is detected, appropriate prevention and control measures could be taken before the disease spreads over a large area.

6.1 Microscopic image changes of wheat blades infected with stripe rust

Three leaf samples were collected and analyzed for obtaining microscopic images in this study. They were the infected blades at 8 days post inoculation (dpi), the diseased blades at 15 dpi, and the healthy blades as the controls. **Figure 15** depicts the reaction of wheat tissue at the different stages of stripe rust development in microscopic sections of primary infection sites. The healthy ones had a complete structure of crosscut sections, a compact structure and plentiful epidermal cells, and the chloroplasts were attached around the cell wall (**Figure 15 A1, A2** and **A3**). When the blades were infected, the fungus would break through the protection of the epidermal cells to destroy the internal structure. Mesophyll cells were modified, and the number of chloroplasts was reduced considerably at 8 dpi (**Figure 15 B1, B2** and **B3**). For the diseased blades (15 dpi), the fungus invaded each tissue of the blade and accumulated to create a large number of summer spore banks or spore beds (**Figure 15 C1, C2** and **C3**).

6.2 Grading method of disease severity of wheat stripe rust using hyperspectral imaging technology

Identification and classification of wheat stripe rust plays an important role in managing this destructive disease to insure productivity of wheat. It helps to

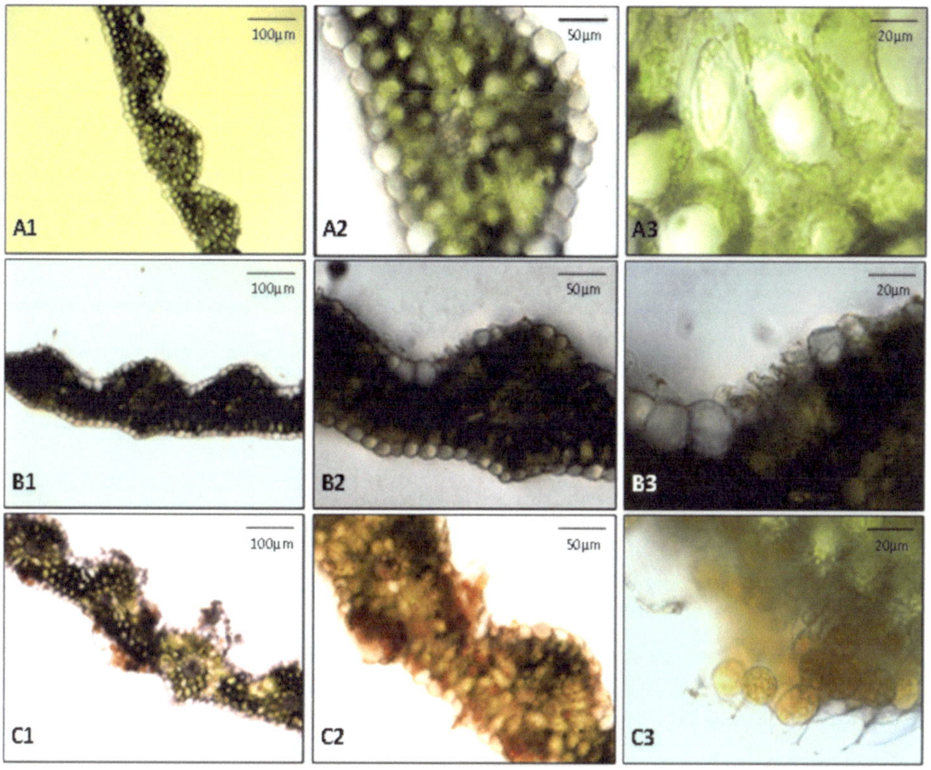

Figure 15.
Sections of wheat leaf tissue infected with Puccinia striiformis f. sp. tritici at different stages of disease development: (A) healthy tissue, (B) tissue at the infection site at 8 days post inoculation (dpi) without macroscopic symptoms but with the epidermis layer destructed, and (C) chlorotic tissue with hyphae of the fungus.

quantitatively assess the level of wheat stripe rust severity in the field to develop strategies to achieve effective control of wheat stripe rust at early stages. Currently, estimation of severity of stripe rust is mainly relied on naked-eye observations. However, these methods are labor-intensive, time-consuming, besides requiring workers with high professional knowledge. In order to quickly and accurately evaluate the disease level of wheat stripe rust, a novel grading method based on hyperspectral imaging technology was proposed to determine severity levels of wheat stripe rust. Firstly, hyperspectral images were captured from 320 infected wheat leaf samples with different levels of disease severity and 40 healthy wheat leaf samples by a HyperSIS hyperspectral system covering the visible and near-infrared region (400 ~ 1000 nm). Secondly, via the analysis of spectral reflectance of leaf and background regions, there were obvious differences in spectral reflectance at the 555 nm wavelength. Therefore, the image of the 555 nm wavelength was named the feature image, which was manipulated by threshold segmentation to obtain a mask image. The logical and operation was conducted by using the original hyperspectral image and mask image to remove the background information. Thirdly, the principal component analysis (PCA) method was used for the dimension reduction of hyperspectral images. The operation results showed that the second principal component image (PC2) can significantly identify the stripe rust-infected area from the healthy area. Based on this, stripe rust-infected area was efficiently segmented by an Otsu method. Finally, the levels of wheat stripe rust severity was graded according to the proportion of stripe rust-infected area on the whole leaf area. To verify the effectiveness of the proposed method, a total of 270

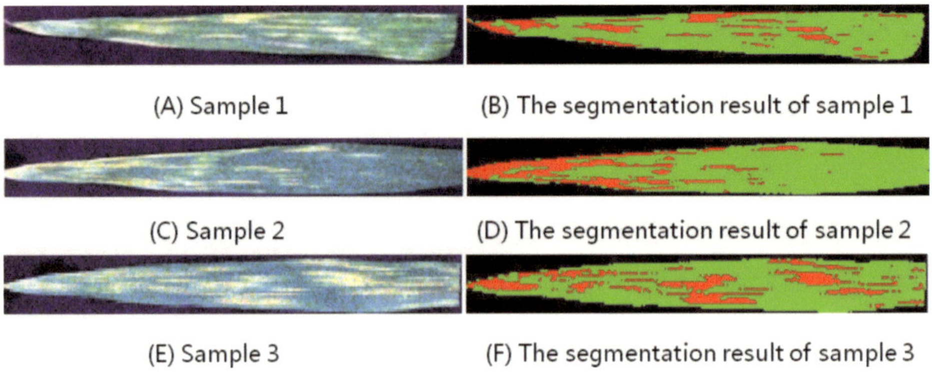

Figure 16.
Processing results of partial disease samples.

leaf samples were collected and evaluated. Experimental results showed that 265 samples could be accurately classified at different disease severities of wheat stripe rust and the overall classification accuracy was 98.15%. The treatment effect of some disease samples was shown in **Figure 16**. The results indicate that hyperspectral imaging technology has the potential to detect stripe rust in wheat.

7. Conclusion and prospects

In this article, we focus on using spectroscopy-based technology to explore developing innovative methods that are nondestructive, rapid, and accurate to diagnose and detect FHB, powdery mildew, and wheat stripe rust, three major fungal diseases in wheat worldwide. Based on their characteristic symptoms and disease development at different growth stages, we have proposed different detection methods and evaluated their performance. The conclusion is that spectroscopy technology can be used as an innovative tool for detection and monitoring of wheat diseases. However, the following limitations and challenges should be considered:

1. The experiments described in the current study are completed in the laboratory. Therefore, the stability and universality of diseased detection models developed from this study need to be further verified. Further field studies should be carried out to validate the model performance on different wheat cultivars and in different cropping areas.

2. Given the high levels of spread and severity the three wheat diseases pose, resulting in very limited window of time for action to control, it is very desirable to develop a detection method that is rapid and can cover a large area at a time. Integration of spectroscopy disease detection with UAV or satellite could be an effective approach toward this goal in the future efforts.

3. The high costs of spectroscopy sensors can also limit the potential application of this spectroscopy technology for disease detection. Future research efforts should be focused in designing a multispectral camera with a low price, improving image data processing speed, and developing methods that can detect multiple diseases simultaneously. In addition, it would be desirable to develop special disease detection device that is combined with the agricultural internet of things (IOT), agricultural situation investigation, and weather stations to detect crop diseases as quickly and accurately as possible, in the

form of multi-point investigation of smartphone terminals and fixed point collection in the frequently-occurring disease areas. These will provide important technical support for timely detection, monitoring, and control of wheat diseases.

Author details

Fenfang Lin[1], Dongyan Zhang[2], Xin-Gen Zhou[3]* and Yu Lei[2]

1 School of Remote Sensing and Geomatics Engineering, Nanjing University of Information Science & Technology, Nanjing, China

2 Anhui Engineering Laboratory of Agro-Ecological Big Data, Anhui University, Hefei, Anhui, China

3 Texas A&M AgriLife Research and Extension Center, Texas A&M University System, Beaumont, Texas, USA

*Address all correspondence to: xzhou@aesrg.tamu.edu

IntechOpen

© 2021 The Author(s). Licensee IntechOpen. This chapter is distributed under the terms of the Creative Commons Attribution License (http://creativecommons.org/licenses/by/3.0), which permits unrestricted use, distribution, and reproduction in any medium, provided the original work is properly cited. CC BY

References

[1] Shewry PR, Hey SJ. The contribution of wheat to human diet and health. Food Energy Secur. 2015; 4:178–202. DOI: https://doi.org/10.1002/fes3.64

[2] FAO. Food and Agriculture Organisation of the United Nations, FAOSTAT, FAO Statistics Division. 2020. Available from http://faostat.fao.org/site/567/default.aspx#ancor

[3] Roser M, Ritchie H, Ortiz-Ospina E. World population growth. Published online at OurWorldInData.org. 2013 Retrieved from: https://ourworldindata.org/world-population-growth

[4] Godfray HCJ, Beddington JR, Crute IR, Haddad L, Lawrence D, Muir JF, Toulmin C. Food security: The challenge of feeding 9 billion people. Science. 2010; 327: 812–818. DOI: 10.1126/science.1185383

[5] Juroszek P, Tiedemann A. Climate change and potential future risks through wheat diseases: a review. Eur. J. Plant Pathol. 2013; 136:21–33. DOI: 10.1007/s10658-012-0144-9

[6] The World Bank. Reducing climate-sensitive risks. 2014. Available at http://documents.worldbank.org/curated/en/486511468167944431/Reducing-climate-sensitive-diseaserisks

[7] Zhang JC, Huang YB, Pu RL, Gonzalez-Moreno P, Yuan L, Wu KH, Huang WJ. Monitoring plant diseases and pests through remote sensing technology: A review. Comput. Electron. Agric. 2019; 165:104943. DOI: 10.1016/j.compag.2019.104943

[8] Karthikeyan L, Chawla I, Mishra AK. A review of remote sensing applications in agriculture for food security: Crop growth and yield, irrigation, and crop losses. J. Hydrol. 2020; 586:124905. DOI: 10.1016/j.jhydrol.2020.124905

[9] Ali MM, Bachik NA, Muhadi N 'A, Yusof TNT, Gomes C. Non-destructive techniques of detecting plant diseases: a review. Physiol. Mol. Plant Pathol. 2019; 108: 101426. DOI: https://doi.org/10.1016/j.pmpp.2019.101426

[10] Khaled AY, Aziz SA, Bejo SK, Nawi NM, Seman IA, Onwude DI. Early detection of diseases in plant tissue using spectroscopy – applications and limitations. Appl. Spectrosc. Rev. 2017; 53: 1–68. DOI: 10.1080/05704928.2017.1352510

[11] Palacios SA, Erazo JG, Ciasca B, Lattanzio VMT, Reynoso MM, Farnochi MC, Torres AM. Occurrence of deoxynivalenol and deoxynivalenol-3-glucoside in durum wheat from Argentina. Food Chem. 2017; 230: 728–734. DOI: https:// doi.org/10.1016/j.foodchem.2017.03.085

[12] Zhang JC, Pu RL, Wang JH, Huang WJ, Yuan L, Luo JH. Detecting powdery mildew of winter wheat using leaf level hyperspectral measurements. Comput. Electron. Agric. 2012; 85:13–23. DOI: https://doi.org/10.1016/j.compag.2012.03.006

[13] Ali S, Leconte M, Rahman H, Saqib MS, Gladieux P, Enjalbert J, De Vallavieille-Pope C. A high virulence and pathotype diversity of Puccinia striiformis f.sp. tritici at its centre of diversity, the Himalayan region of Pakistan. Eur. J. Plant Pathol. 2014a; 140: 275–290. DOI: https://doi.org/10.1007/s10658-014-0461-2

[14] Ali S, Gladieux P, Leconte M, Gautier A, Justesen AF, Hovmøller MS, De Vallavieille-Pope C. Origin, migration routes and worldwide population genetic structure of the wheat yellow rust pathogen Puccinia striiformis f.sp. tritici. PLoS Pathog. 2014b; 10: e1003903. DOI: https://doi.org/10.1371/journal.ppat.1003903

[15] Atta BM, Saleem M, Ali H, Bilal M, Fayyaz M. Application of fluorescence spectroscopy in wheat crop: early disease detection and associated molecular changes. J. Fluoresc. 2020; 30: 1–10. DOI: https://doi.org/10.1007/s10895-020-02561-8

[16] Bürling K, Hunsche M, Noga G, Pfeifer L, Damerow L. UV-induced fluorescence spectra and lifetime determination for detection of leaf rust (*Puccinia triticina*) in susceptible and resistant wheat (*Triticum aestivum*) cultivars. Funct. Plant Biol. 2011; 38: 337–345. DOI: 10.1071/fp10171

[17] Lüdeker W, Dahn HG, Günther KP. Detection of Fungal Infection of Plants by Laser-induced Fluorescence: An Attempt to Use Remote Sensing. J. Plant Physiol. 1996; 148: 579–585. DOI: https://doi.org/10.1016/S0176-1617(96)80078-2

[18] Tischler YK, Thiessen E, Hartung E. Early optical detection of infection with brown rust in winter wheat by chlorophyll fluorescence excitation spectra. Comput. Electron. Agric. 2018; 146: 77–85. DOI: https://doi.org/10.1016/j.compag.2018.01.026

[19] Padhye P, Rajani K. Machine vision guided system for classification and detection of plant diseases using support vector machine. Int. J. Electron. Commun. Comput. Eng. 2014; 5: 249–254.

[20] Padmavathi K, Thangadurai K. Identification of plant leaves disease detection and optimal solution using genetic algorithm. Int. J. Adv. Res. Comput. Sci. Softw. Eng. 2015; 5: 1165–1168.

[21] Johannes A, Picon A, Alvarez-Gila A, Echazarra J, Rodriguez-Vaamonde S, Navajas AD, Barredo AO. Automatic plant disease diagnosis using mobile capture devices, applied on a wheat use case. Comput. Electron. Agric. 2017; 138: 200–209. DOI: https://doi.org/10.1016/j.compag.2017.04.013

[22] Wang H, Li G, Ma Z, Li X. Image recognition of plant diseases based on backpropagation networks. In: 2012 5th Int Congr Image Signal Process. 2012. p. 894–900. DOI: https://doi.org/10.1109/CISP.2012.6469998

[23] Ngugi LC, Abelwahab M, Abo-Zahhad M. Recent advances in image processing techniques for automated leaf pest and disease recognition - a review. Information Processing in Agriculture. DOI: https://doi.org/10.1016/j.inpa.2020.04.004

[24] Ferentinos KP. Deep learning models for plant disease detection and diagnosis. Comput. Electron. Agric. 2018; 145:311–318. DOI: https://doi.org/10.1016/j.compag.2018.01.009

[25] Liu Z, Wu H, Huang J. Application of neural networks to discriminate fungal infection levels in rice panicles using hyperspectral reflectance and principal components analysis. Comput. Electron. Agric. 2010; 72: 99–106. DOI: https://doi.org/10.1016/j.compag.2010.03.003

[26] Yuan L, Zhang J, Shi Y, Nie C, Wei L, Wang J. Damage mapping of powdery mildew in winter wheat with high-resolution satellite image. Remote Sens. 2014b; 6:3611–3623. DOI: 10.3390/rs6053611

[27] Huang W, Lamb DW, Niu Z, Zhang Y, Liu L, Wang J. Identification of yellow rust in wheat using in-situ spectral reflectance measurements and airborne hyperspectral imaging. Precis. Agric.2007; 8: 187–197. DOI: 10.1007/s11119-007-9038-9

[28] Zhang J, Pu R, Huang W, Yuan L, Luo J, Wang J. Using in-situ hyperspectral data for detecting and discriminating yellow rust disease from nutrient stresses. Field Crop. Res. 2012a;

134: 165–174. DOI: https://doi.org/10.1016/j.fcr.2012.05.011

[29] Bürling K, Hunsche M, Noga G. Presymptomatic detection of powdery mildew infection in winter wheat cultivars by laser-induced fluorescence. Appl. Spectrosc. 2012; 66: 1411–1419.

[30] Römer C, Bürling K, Hunsche M, Rumpf T, Noga G, Plümer L. Robust fitting of fluorescence spectra for pre-symptomatic wheat leaf rust detection with Support Vector Machines. Comput. Electron. Agric. 2011; 79: 180–188. DOI: https://doi.org/10.1016/j.compag.2011.09.011

[31] Iqbal MJ, Goodwin PH, Leonardos ED, Grodzinski B. Spatial and temporal changes in chlorophyll fluorescence images of Nicotiana benthamiana leaves following inoculation with Pseudomonas syringae pv. tabaci. Plant Pathol. 2012; 61: 1052–1062. DOI: 10.1111/j.1365-3059.2012.02592.x

[32] Scholes JD, Rolfe SA. Chlorophyll fluorescence imaging as tool for understanding the impact of fungal diseases on plant performance: a phenomics perspective. Fuct. Plant Biol. 2009; 36: 880–892.

[33] Barbedo JG, Tibola CS, Fernandes JM. Detecting Fusarium head blight in wheat kernels using hyperspectral imaging. Biosyst. Eng. 2015; 131: 65–76. DOI: https://doi.org/10.1016/j.biosystemseng.2015.01.003

[34] Ravikanth L, Singh CB, Jayas DS, White ND. Classification of contaminants from wheat using near-infrared hyperspectral imaging. Biosyst. Eng. 2015; 135: 73–86. DOI: https://doi.org/10.1016/j.biosystemseng.2015.04.007

[35] Weber VS, Araus JL, Cairns JE, Sanchez C, Melchinger AE, Orsini E. Prediction of grain yield using reflectance spectra of canopy and leaves in maize plants grown under different water regimes. Field Crop Res. 2012; 128: 82–90. DOI: 10.1016/j.fcr.2011.12.016

[36] Huang WJ, Guan QS, Luo JH, Zhang JC, Zhao JL, Liang D, Huang LS, Zhang DY. New optimized spectral indices for identifying and monitoring winter wheat diseases. IEEE J. Sel. Top. Appl. Earth Obs. Remote Sens. 2014; 7: 2516–2524. DOI: 10.1109/JSTARS.2013.2294961

[37] Mahlein AK, Rumpf T, Welke P, Dehne HW, Plümer L, Steiner U, Oerke EC. Development of spectral indices for detecting and identifying plant diseases. Remote Sens. Environ. 2013; 128: 21–30. DOI: https://doi.org/10.1016/j.rse.2012.09.019

[38] Zhang D, Wang Q, Lin F, Yin X, Gu C, Qiao H. Development and evaluation of a new spectral disease index to detect wheat fusarium head blight using hyperspectral imaging. Sensors. 2020a; 20:2260. DOI: 10.3390/s20082260

[39] Pourreza A, Pourreza H, Abbaspour-Fard MH, Sadrnia H. Identification of nine Iranian wheat seed varieties by textural analysis with image processing. Comput. Electron. Agric. 2012;83: 102–108. DOI: 10.1016/j.compag.2012.02.005

[40] Kingsbury N. Complex wavelets for shift invariant analysis and filtering of signals. Appl. Comput. Harmon. A. 2001; 10: 234–253. DOI: 10.1006/acha.2000.0343

[41] Selesnick IW, Baraniuk RG, Kingsbury NG. The dual-tree complex wavelet transform. IEEE Signal Proc. Mag. 2005; 22: 123–151. DOI: 10.1109/MSP.2005.1550194

[42] Yang P, Zhang FL, Yang GW. Fusing DTCWT and LBP based features

for rotation, illumination and scale invariant texture classification. IEEE Access. 2018; 13336–13349. DOI: 10.1109/ACCESS.2018.2797072

[43] Li TS, Wang J, Tu MS, Zhang Y, Yan YH. Enhancing link prediction using gradient boosting features. Intelligent Computing Theories and Application, Icic 2016. 2016; 9772: 81–92. DOI: 10.1007/978-3-319-42294-7_7

[44] Zhang D, Chen G, Yin X, Hu R, Gu C, Pan Z, Zhou X, Chen Y. Integrating spectral and image data to detect Fusarium head blight of wheat. Comput. Electron. Agric. 2020b; 175: 105588. DOI: https://doi.org/10.1016/j.compag.2020.105588

[45] Sankaran S, Mishra A, Ehsani R, Davis C. A review of advanced techniques for detecting plant diseases. Comput. Electron. Agric. 2010; 1: 1–13.

[46] Jafari M, Minaei S, Safaie N, Torkamani-Azar F. Early detection and classification of powdery mildew-infected rose leaves using ANFIS based on extracted features of thermal images. Infrared Phys. Technol. 2016; 76: 338–345.

[47] Lin FF, Wang DD, Zhang DY, Yang XD, Yin X, Wang DY. Evaluation of spectral disease index PMI to detect early wheat powdery mildew using hyperspectral imagery data. Int. J. Agric. Biol. 2018; 20: 1970–1978. DOI: 10.17957/IJAB/15.0716

[48] Huang LS, Zhang Q, Zhang DY, Lin FF, Chao X, Zhao JL. Early diagnosis of wheat powdery mildew using Relief-F band screening. Infrared and Laser Engineering. 2018; 47: 210–217.

[49] Pu R, Kelly M, Chen Q, Gong P. 2008. Spectroscopic determination of health levels of coast live oak (*Quercus agrifolia*) leaves. Geocarto Int. 2008; 23: 3–20.

[50] Kira K, Rendell L. A practical approach to feature selection. Proc. 9th Int. Workshop Mach. Learn., San Mateo, CA, USA. Morgan Kaufmann Publishers Inc; 1992. P. 249–256.

[51] Robnik-Sikonja M, Kononenko I. Theoretical and empirical analysis of ReliefF and RReliefF. Mach. Learn. 2003; 53: 23–69.

[52] Huang LS, Zhang DY, Liang D, Yuan L, Zhao JL, Hu GS, Du SZ Xu XG. Continuous wavelet analysis for diagnosing stress characteristics of leaf powdery mildew. Int. J. Agric. Biol. 2013; 15: 34–40. DOI: 10.1684/agr.2012.0578

[53] Thenkabail PS, Smith RB, De Pauw E. Hyperspectral vegetation indices and their relationships with agricultural crop characteristics. Remote Sens. Environ. 2000; 2: 158–182. DOI: https://doi.org/10.1016/S0034-4257(99)00067-X

[54] Filella I, Serrano L, Serra J, Penuelas J. Evaluating wheat nitrogen status with canopy reflectance indices and discriminant analysis. Crop Sci. 1995; 5: 1400–1405.

[55] Broge NH, Leblanc E. Comparing prediction power and stability of broadband and hyperspectral vegetation indices for estimation of green leaf area index and canopy chlorophyll density. Remote Sens. Environ. 2001; 2: 156–172. DOI: https://doi.org/10.1016/S0034-4257(00)00197-8

[56] Gamon JA, Penuelas J, Field CB. A narrow-waveband spectral index that tracks diurnal changes in photosynthetic efficiency. Remote Sens. Environ. 1992; 1: 35–44. DOI: https://doi.org/10.1016/0034-4257(92)90059-S

[57] Kim MS, Daughtry CST, Chappelle EW, McMurtrey JE. The use of high spectral resolution bands for estimating absorbed photosynthetically

active radiation (APAR). Proc. 6th Int. Symposium on Physical Measurements and Signatures in Remote Sensing. France Val d'Isere; 1994. P. 299–306.

[58] Haboudane D, Miller JR, Pattery E, Zarco-Tejad PJ, Strachan IB. Hyperspectral vegetation indices and novel algorithms for predicting green LAI of crop canopies: Modeling and validation in the context of precision agriculture. Remote Sens. Environ. 2004; 3: 337–352. DOI: https://doi.org/10.1016/j.rse.2003.12.013

[59] Daughtry CS, Walthall CL, Kim MS, de Colstoun EB, McMurtrey JE. Estimating corn leaf chlorophyll concentration from leaf and canopy reflectance. 2000. Remote Sens. Environ. 2000; 2: 229–239. DOI: https://doi.org/10.1016/S0034-4257(00)00113-9

[60] Merton R, Huntington J. Early simulation of the ARIES-1 satellite sensor for multi-temporal vegetation research derived from AVIRIS. Summaries of the Eight JPL Airborne Earth Science Workshop, Pasadena, CA: JPL, Publication; 1999. P. 299–307.

[61] Merzlyak MN, Gitelson AA, Chivkunova OB, Rakitin VY. Non-destructive optical detection of pigment changes during leaf senescence and fruit ripening. Physiol. Plant. 1999; 1: 135–14. DOI: 10.1034/j.1399-3054.1999.106119.x

[62] Gitelson A, Merzlyak MN, Chivkunova OB. Optical properties and nondestructive estimation of anthocyanin content in plant leaves. Photochem. Photobiol. 2001; 1: 38–45. DOI: 10.1562/0031-8655(2001)074<0038:opaneo>2.0.co;2

Chapter 6

The Trends in the Evaluation of Fusarium Wilt of Chickpea

Chandan Singh and Deepak Vyas

Abstract

Chickpea (*Cicer arietinum* L.) is one of the important annual legume crops, cultivated throughout the India since ancient time. It is also grown in many countries of the world. The crop has been facing numerous biotic and abiotic constraints. Among biotic constraint crop affected adversely by diseases, caused by many pathogens. Ever since 1918 when for the first time wilt disease of chickpea was reported and *Fusarium oxysporum* f. sp. *ciceri* was the causal organism many strategies have been adopted to control the wilt disease. The controlling methods included conventional as well as modern one. However, more and more emphasis was given on biological control agents such as AM fungi and *Trichoderma*. The role of AM fungi have been evaluated for controlling the wilt disease similarly role of *Trichoderma* is thoroughly established biological control agent against Fusarium wilt. With the advent of modern tools and techniques developing markers, resistant varieties, all such sources enable us to reduce the effect of pathogens. Here an attempted has been made to acknowledge the trend of disease management and evaluation strategies of Fusarium wilt of chickpea for getting better yields of the crop.

Keywords: pathogenesis, pathotype, ITS, PCR-based diagnosis, chickpea

1. Introduction

Chickpea (*Cicer arietinum* L.) is one of the major legume crops grown in the cool season, mostly in dryland [1]. India is the largest chickpea producer as well as consumers in the world [2]. Chickpea is also known as garbanzo bean or Bengal gram, is a self-pollinated, annual diploid (2n = 2x = 16) species with a genome size of 738 Mb [3] belongs to the family of Fabaceae (Leguminoseae) sub-family Faboideae (*Papilionaceae*) and tribe *Cicereae* it has 9 annual and about 34 perennial wild species. Among 9 annual species, chickpea is the only cultivated species worldwide [4, 5]. Chickpea served as a major dietary protein for humans at the cheapest cost compare to other sources of protein. Like other crops, chickpea is subjected to many abiotic and biotic stress, which, causes limited production as per the theoretical potential. The production of chickpea in the Indian subcontinent and Asian countries are severely affected by the pathogenic fungus, bacteria, virus, and nematodes which causes disease like *Fusarium* wilt, dry root rot, *Ascochyta* blight, Collar rot, Bacterial blight, Filiform virus and Root nematode [2, 6]. The wilt disease caused by FOC was more hazardous and the impact was so severe, that sometimes no yield is recorded, however, the percentage of loss depends upon agroclimatic conditions of the region [7]. FOC is a soil-borne disease and the mode of dissemination are many such as infected plant debris (root, leaf, and stem), soil and seed as mycelium, microconidia, macroconidia, and, most commonly, as chlamydospores [8–10]. Pathogenic variability has been observed on the basis of symptoms causes to the host, FOC subdivided into two pathotypes i.e. wilting pathotype and yellowing pathotype, FOC has eight races 0, 1A, 1B/C, 2, 3, 4, 5, and 6 that have been identified till now [8, 10–13]. The Races 0 and 1B/C are of the yellowing pathotypes, while races 1A and 2,3,4,5, and 6 belong to the wilting pathotypes, and caters are known for their economic loss to the farmer [14]. The variability in the pathogenicity makes it more complex and difficult to identify them. Therefore the proper identification of the pathogens is the most important part of the management of the disease and in developing resistance variety of the crop against the pathogen. However, to identify the causal organism applying molecular tools such as PCR techniques, specific primers, and PCR assays have been developed [9, 11, 12, 15, 16]. The evaluation of *Fusarium* wilt is involved in many stages, which has been discussed in this chapter and also will list out the troubleshoots in the diagnosis and the identification protocols both morphological and molecular method.

2. Pathogenesis and pathophysiology

The FOC shows variableness inside their populations also based on their severity index their virulence is decided, it was discussed in the above sections that FOC has eight pathotypes i.e. yellow pathotypes and wilting pathotypes. The yellow pathotypes races cause progressive foliar yellowing accompanied by vascular discoloration and late plant death while the wilting pathotypes cause fast and extreme chlorosis flaccidity, vascular discoloration, and early plant death, the life cycle of the pathogen is of two-phase; the parasitic phase and the saprophytic phase (**Figure 1**; [17]). At the parasitic phase, the fungus gets to enter or invade the host through penetration by crack, wounds, roots hairs root cap, or root branches, the penetration process is likely to be enhanced by the hydrolytic enzymes, after the establishment of the pathogen into the roots, the cortical region is colonized by the emerging mycelia and slowly and gradually enters into the vascular stream and distributed throughout the plants and the gradual blockage of the vessel happens by the mycelium and the spores that were reflected by the appearance of the symptoms on the host [18–20]. It does not matter how the infected plant died either by wilting

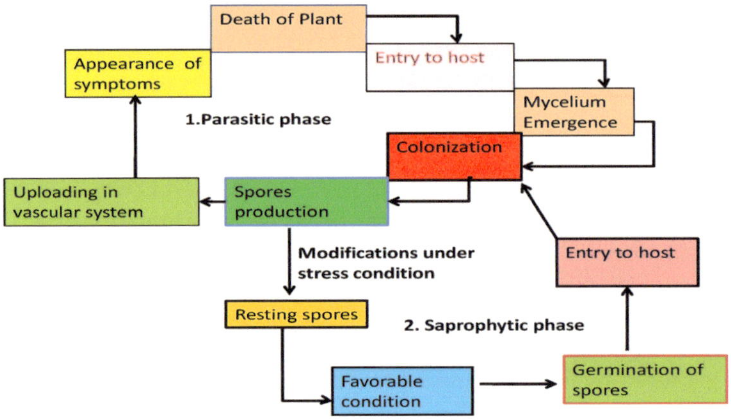

Figure 1.
Illustration showing the parasitic and saprophytic phase of Fusarium oxysporum *f. sp.* ciceri *on Chickpea.*

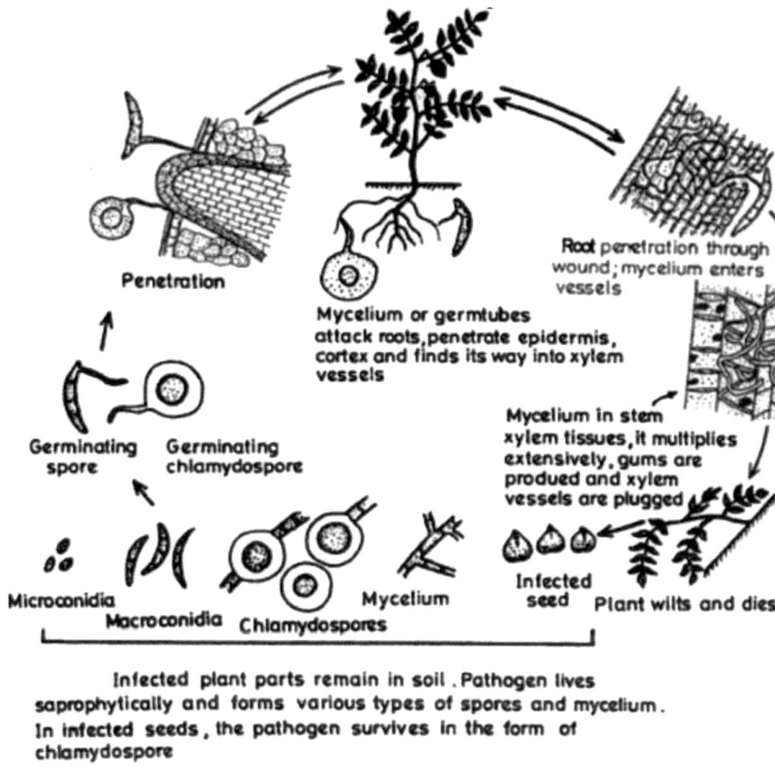

Figure 2.
Disease cycle of F. oxysporum *f. sp.* ciceris *(source: Jalali and Chand, 1992) [21].*

or chlorosis the important thing is that the pathogen survive through resting spores, the importance of the present monograph lies in how *Fusarium* causes wilt disease in chickpea and what measure has been used to control the disease is discussed in length (**Figure 2**).

3. Pathogenic variability in *Fusarium. oxysporum* f. sp. *ciceris*

The FOC shows high variability with respect to its morphology (**Figure 3A** and **B**), cultural characteristics, and virulence, In the number of studies that have been carried out suggested high pathogenic variability among the isolates of different regions [22]. In Mexico PCR based study on 355 isolates reported that 161 strain are positive for the FOC and has shown morphological variability in FOCs and also revealed that it is not a function of the physical and chemical properties of the soil, nor the geographic location of the crop fields rather it is a race of the FOC [23]. In a laboratory study carried out in Bangladesh, FOC isolates showed significant variability in their cultural, morphological, and physiological traits, i.e. colony color, shape, margin, and texture; mycelial radial growth, and spore production [24] considering the above facts it can be established that FOC has significant variability despite of their occurrence. Variance in the isolates of FOC has led to the designation of pathogenic races, and the different races of FOC can be identified by their severity on a different set of differential chickpea cultivars [13]. Pathogenic races of FOC differ in geographic distribution [17]. Races 0, 1B/C, 5, and 6 are found mainly in the Mediterranean Basin and California (USA) Race 1A has been reported in India, United States, and the Mediterranean regions [13, 19, 25], whereas races 2 and 3 have been reported in

Figure 3.
Showing yellowing symptom (a) and wilting symptom (B).

Ethiopia, India, and Turkey and race 4 has been reported in Ethiopia, India, and Iraq [13, 26]. In short, it can be said that FOC showed dynamic variability respective to geographical distribution. One must consider races for the evaluation and management of diseases. Some diagnostic trends have been discussed here.

4. Pathogenicity test for FOC

Confirmation of the pathogen is commonly conducted by the methods of Koch Postulations, for pathogenicity test in the chickpea caused by the FOC are added here:

a. Blotter paper technique:

In this method, the Pure culture of pathogens is obtained from the infected host, the obtained fungus is then multiplied in potato-dextrose broth and after incubation, the mycelium mat were collected and macerated in distilled water then stored, simultaneously the host seedling of 5 days old grown on riverbed sand soil was uprooted, the roots were washed with running tap water and these roots were dipped in the solution of inoculum by touching the edge of the beaker then after the seedling of the test line are placed side by side on a blotter paper (size 45 cm × 25 cm with one fold.) so that the roots are covered and the green part remains outside of this paper. Folded blotters were kept, one on top of the other, in heaps of 10 in a tray. The trays were placed in an incubator at 25° C with 12-hr artificial light. After 8 days of incubation, blotters were moistened adequately every day. The seedlings were examined for the extent of root damage and a score of the disease [27]. Nene et al. [28] also recommended this technique for screening of chickpea resistant variety against FOC.

b. Soil Inoculation method/Pot Screening:

To find out the variability among the different isolates of pathogens, the investigation on soil inoculation by FOC is a common method. In the soil inoculation method, different isolates of FOC were multiplied individually on sand and soil (3:1) medium and filled in plastic pots. The high inoculum level of 20% w/w is used for the maximum wilting. The inoculated and un-inoculated control pots are maintained under controlled environmental conditions. Observations on wilt incidence in seedlings were recorded from 10 days after sowing (DAS) to 30 DAS.

c. Sick plot method:

The sick plot method is a method used to check the pathogenicity and screening of chickpea cultivar lines against FOC. This method is conducted in field conditions. A suitable size plot is developed into a sick plot, about six-years required to develop a sick plot by the pathogen inoculums [28] following steps are given below.

1. Select a plot of adequate size and ensure that it is isolated from other chickpea fields to avoid the spread of the fungus inoculums from this plot to others. The selected plot should have been harvested in the previous year with chickpea, and at least traces of wilt incidence should have been observed.

2. Wilted plants from other fields were collected in bulk, chop into small pieces, and incorporate uniformly in the surface soil of the plot.

3. Susceptible cultivars were grown in the selected plot. Make sure a good plant population and maintain normal agronomic operations.

4. At the end of the season, the cultivar shows 20% wilting symptoms, then all the debris of the cultivar were scattered after harvesting and threshing uniformly all over the plot and incorporate it by discing. Step II may be repeated this will help in increasing the level of the inoculums to make the soil "sick."

5. Steps III and IV were repeated in the next season. More than 90% wilt incidence in the selected plot should be observed. If the incidence is less than 70%, then steps III and IV are repeated.

6. Screening should be initiated in the next season. Sow a susceptible cultivar after every two test rows in the whole field, these rows will serve to checks in monitoring and maintaining the wilt sickness of the plot. Now the susceptible check rows must show more than 90% wilt.

7. From the 4th or 5th year onwards, plant every fifth row as a susceptible check. This will provide for more breeding material and at the same time maintain the level of sickness.

8. Planting any other crop in this plot is not recommended.

Sometimes even taking all the care, a sick plot which is predominated by FOC also provides room for other soil-borne pathogens.

5. Isolation and maintenance of pathogens

The pathogens can be isolated either from an infected host or from the soil. To isolate the pathogen from the infected plant parts, the plant sample collected randomly, basically, the root and stem of the plant is equally used to isolate the pathogen. The obtained plant sample with 1:1000 mercuric chloride solutions for one minute and washed twice with sterile water after cleaning the roots or stem. Cut the section of the root or the stem then placed it on the PDA medium and incubate at 28-30°C to get the fungal colony. The colony obtained on the Petri plate of PDA medium is then purified by using the single spore culture method in PDA slant. If the pathogen has to be isolated from soil then the soil sample must be obtained from the infected crop field and using serial dilution methods, the culture is made on the Petri Plate of PDA (10^{-6}–10^{-8} dilution is appropriate dilution for the culture) to obtain the colony followed by the single spore isolation method in slant for pure culture. The fungus slant can be store for further study.

6. Diagnostic analysis of FOC

Diagnosis is an integral part to analyze any disease and the approach to establish host and pathogen interactions. The approach could be conventional or advanced. Methods have been discussed in details see below.

6.1 Diagnosis at morphological and at the anatomical level

The morphological study includes observation of the cultivar in the field and the symptom appearance; it was discussed in the section (1.2) that the FOC consist of two pathotypes i.e. yellow pathotypes causing yellowing of the foliar (**Figure 3A**) and gradually the death of the plants and other is wilting (**Figure 3B**) where the chickpea infested by the FOC wilted and the cultivar dies faster and led to the destruction of the crops. The chickpea infected by FOC has a formation of root nodule or has a trace of root nodule (**Figure 4A** and **B**.).

Under anatomical studies, the fungal mycelium present in the vessel lumen of infected crops, entry of mycelium follows an intracellular path to the cortex and enters xylem vessels through the pits, the mycelium in xylem branches and produce microconidia, these microconidia detached and carried upward by the vascular system and the lateral movements of microconidia between the vessels are through the pit, as a result, the water economy of infected plants is eventually severely compromised by blockage of vessels, resulting in stomatal closure, wilting and death of leaves, often followed by the death of the whole plant [9, 25]. *Fusarium oxysporum* f. sp. *ciceris* can survive as mycelium and chlamydospores in seed and soil, and also on infected crop residues, roots and stem tissue buried in the soil for up to six years [5, 20, 29]. Different isolates of FOC have cultural variation, when isolates of FOC grown on PDA shows fluffy, partial fluffy, and white cottony mycelium growth having different pigmentations like

Figure 4.
Figure showing the root nodulation in infected and healthy chickpea; A- No or trace nodule, B- root nodule in health chickpea.

gray purple, orange group, and gray-brown [5], one must undergo an in-vitro study for the details cultural description of the FOC since it has a high rate of variability.

6.2 Diagnosis at the molecular level

The identification of the Pathogenic race of FOC is predominantly based on the differential reactions on the host genotypes they infected [30]. Soon after the development of PCR and other biotechnological approaches like Random Amplified polymorphic DNA (RAPD), Amplified fragment Length Polymorphism (AFLP), Simple sequence repeat (SSR), and inter-simple sequence repeat (ISSR) are being used to study the pathogenic race and the pathogenic variability [30]. To study the pathogen colonization and progression in wilt susceptible and resistant variety of the chickpea, insertion of GPF gene in the pathogenic race using the molecular tool helps to visualize the colonization and progression under confocal microscopy, moreover, quantitative PCR (qPCR) is helpful in estimation of the pathogen load and the progression of the diseases across various tissue in both the cultivar during the course of the disease [31]. The synergistic gene-specific markers, ITS-RFLP, ISSR, and AFLP are used in distinguishing *F. oxysporum* f. sp. *ciceris* races, the use of oligonucleotides designed from the conserved region of Hop78 transposon helps in distinguishing the FOC races 1,2 and 4 from race 3 by PCR amplification methods [32]. Moreover, the genetic variability of different isolates of FOC race can also be established by using the molecular marker, RAPDs, and AFLP, and help in identifying the different clusters of the gene that represent the races of the pathogens from the coordinate analysis of the similarity index data generated from the molecular marker studies [33]. The population of pathogens is genetically as well as pathologically distinct from those in other countries which can be reported from DNA fingerprinting studies [34].

7. Development of *Fusarium* wilts in relation to environment

Agro-climatic factors are solely related to crop yield and the virulence of plant pathogens. Changes in climatic factors like temperature, carbon dioxide, and the frequency and intensity of extreme weather could have a significant impact on crop yield. The effects of increased temperature on any crops depend on the crop's optimal temperature for growth and reproduction. Warmer and moist weather favor many pests, weeds, and fungi to thrive. The climatic change affects the multitrophic interaction in the soil, which is primacy for the plants and other functional domains of the soil ecosystem; the soil food web is highly interlinked to all the functional domains of the soil [35]. Soil food web sequesters carbon, helps in cycle nutrients, maintain soil health to suppress pathogens, helps plant tolerates abiotic and biotic stress, and maintain ecosystem resilience and sustainability. Temperature plays an important role in the resistance of the chickpea against the *Fusarium* wilt, the effect of temperature on-resistance of the chickpea cultivars to *Fusarium* wilt that caused by various races of *Fusarium oxysporum f.sp. ciceris* (FOC) when temperature increased by 2-3°C from the normal temperature, the cultivar became more susceptible to pathogens [36]. To combat the effect of the global change in temperature on chickpea production different workers are evaluating for the validation of the chickpea resistance in different environmental constraints. Now to mitigate the different environmental constraints a new approach or model is the demand of today agriculture practices, Anuga and Gordon [37] in this regard has thrown light over the Climate-Smart (CS) weather practices/strategies to mitigate climate-induced agriculture impacts but it would be very imperative to understand the CS weather practices that are effective in influencing farmers productivity positively.

8. Evaluation methods of fungicidal/bioagents on FOC

The antifungal substances obtain the substance that inhibits the growth of the fungus or completely restrict fungal growth. The antifungal activities are done basically *in vitro* or *in vivo* conditions. The methods are described below.

a) Cavity slide test method

The effect of the antifungal substance on the spore germination of the pathogens conducted by using the cavity slide method, For this the spore suspension at the density 10^6 spores/mL was placed on the cavity slide along with the antifungal substance and incubated under mist-chamber at 28–30^0 C. The percentage rate of the spore germination is recorded after 24 hours under a light microscope.

b) Poisoned Food Technique

Food poisoned techniques represent the method of poisoning fungal growth medium by using the antifungal substance or the agent. The reduction of the fungal growth on the medium was recorded, the reduction of the fungal growth indicates that the inhibition was caused by the antifungal substances. For this, the PDA medium is prepared and amended with the desired concentration of the substances, and the mycelial discs of the FOC are placed at the center of the Petri Plate and then incubated at 28-30°C for ten days and the radial mycelium growth are measured. The percent reduction of mycelial growth over control is calculated employing the following formula:

$$\text{Percent decrease over control} = \frac{\text{Dc} - \text{Dt}}{\text{Dc}} \times 100 \qquad (1)$$

where Dc - Average diameter of fungal growth in control; Dt - Average diameter of fungal growth in treatment.

c) Well diffusion method

The well diffusion method is employed to assess the antimicrobic ability of the substance against the FOC, PDA medium was prepared on plates and the sub-cultured pathogen on sterile Potato dextrose broth is swabbed on the plates, then after using a sterile cork borer well are cuts, and are filled with the antifungal substance. The FOC pathogen is allowed to grow under optimal states with a positive and negative control. The diameter of the inhibition zone is recorded for the fungicidal substance.

d) Blotter method

This method is suitable to check the infection rate in the seeds of chickpea by the FOC pathogen, in this method the spore suspension is made and the required number of the seeds are soaked in the fungal spore suspension, and then after seeds are submerged in the fungicidal substance of desired concentration. The treated seeds are placed on the moist blotter paper in a Petri plate and incubated (20–22°C) for seven days, the seeds are examined for the growth of seed-borne pathogens. The result is expressed in the percentage of disease reduction as compared to control.

9. Differences between *Fusarium* and *Verticillium* wilt symptoms

To conduct any successful evaluation of a disease, the critical analysis of the symptom represents a crucial part, therefore the *Fusarium* wilt of chickpea, are frequently confused with *Verticillium* wilt so the difference between them must be highlighted to resolve the confusion between the symptoms they caused, *F. oxysporum* causes vascular discoloration (brown staining of stem tissue) which is darker and continuous whereas *Verticillium* causes more of a "flecking" or "spotty" type of strain. The vascular discoloration in plants infected by *Fusarium* is low in the lower part of the stem than the plant infected with *Verticillium* (more in the lower stem, below the cotyledonary node and upper taproot in *Fusarium*). Another most interesting difference fact is that *Fusarium* wilt is favored by warm temperatures while *Verticillium* wilt is favored by cool temperatures it can be stated that the *Fusarium* became more active in warmer weather and *Verticillium* in the cooler weather [1, 17].

10. Managements of fusarium wilt of Chickpea

A successful evaluation of disease is always followed by management strategies to combat the disease. Management of *Fusarium* wilt of chickpea is difficult to achieve and no single control measure is sufficiently effective [36]. *Fusarium* wilt of chickpea is a monocyclic disease in which development is driven by the pathogen's primary inoculums [13]. *Fusarium* wilt is difficult to control with cultural practices and chemical control is one of the most used approaches. Biological control and plant resistance, however, provide an environmentally and economically appropriate means for disease control that can be easily included within an integrated disease management strategy [38]. In fact, the use of natural resistance for the management of fungal diseases in chickpea may be enhanced through biological control using either bacterial or fungal antagonists [7, 39–42]. The use of arbuscular mycorrhizal consortium to control biologically *Fusarium* wilt is found to be an effective measure to manage wilt disease of chickpea [7, 43]. Different biocontrol agents, including bacteria belonging to the genera *Bacillus*, *Pseudomonas,* and *Rhizobium* and fungi such as nonpathogenic and non-host *Fusarium* species, have been used successfully and resulted in a significant reduction in both pathogenic fungal growth in vitro and disease development in the plant [44–47]. Biofumigation using brassica may be included in alternative strategies for the management of *Fusarium* wilt of chickpea [48]. Induced resistance through the accumulation of various phenolic compounds and phytoalexins, as well as the activation of peroxidases, polyphenoloxidases, and key enzymes in phenylpropanoid and isoflavonoid pathways, play a crucial role in the biological control and resistance of chickpea to pathogenic attacks [49–52]. For instance, recent investigations revealed that the pre-treatment of chickpea seedlings with selected *Rhizobium* isolates before challenging with FOC, increased significantly the levels of total phenolics and the constitutive isoflavonoids, formononetin, and biochanin A [46]. Protection of chickpea against *Fusarium* wilt by the nonpathogenic and non-host *Fusarium* species shown to be associated with the induction of the synthesis of the phytoalexins medicarpin and maachiain and the related isoflavones formononetin and biochanin A [53]. Maakiain and medicarpin exhibited potent antifungal activity towards *Fusarium* spores, by inhibiting their germination and hyphal growth. Studies on phytoalexin tolerance in chickpea pathogenic fungi have also shown a relationship between virulence and the ability of these fungi to detoxify phytoalexins [54]. This can be exemplified by the fact that the fungus *Nectria heamatococca* can metabolize and detoxify maakiain and medicarpin and that these reactions are required for pathogenesis by this fungus on chickpea. Among the other phenolic compounds studied, gallic, cinnamic, ferulic, and chlorogenic acids are also associated with the protection of chickpea from fungal attacks through induced resistance. Of the induced defenses, phenolics and phytoalexin production have received particular attention in chickpea. However, efforts must be made in providing good evidence that these compounds accumulate at the right time, concentration, and location and to elucidate the regulatory genes involved in their rapid and coordinated induction in response to fungal attack [53].

10.1 Use of resistant cultivars

The use of resistant cultivar endure the most convenient and economically efficient way to control measures for the administration of Fusarium wilt of chickpea. Resistant cultivars of chickpea display the key component in integrated disease management programs. Resistance to FOC races has been only recognized in the desi germplasm and lesser extent in Kabuli chickpea as well as in some wild *Cicer spp*. The use of resistant cultivars receives some critical issues due to some undesirable agronomic characteristics. Furthermore, the high pathogenic variability in FOC populations limits the effectiveness and extensive use of available resistance variety [13].

10.2 AMF based fusarium wilt control

The rising concern for community health and the environment due to use of chemicals has fulled the investigation on the biological control methods. The key to achieving successful biological control depends on the knowledge of ecological interaction taking place in the soil and the root atmosphere. AMF fungi are one of the potential biological agents found to possess significant control over the FOC. The introduction of AMF to chickpea suppresses the effects of pathogens in rhizoplane as well as in the rhizosphere [7]. AMF helps in the nodule formation, Species of AMF like, *Glomus ambisporum*, *G. mosseae*, *Acaulospora nicolsonii*, *A. spinosa*, *Glomus fasciculatum,* etc., is reported as best phytoprotection. The percentage of root colonization by AMF depends on different factors, AMF is a strong candidate for controlling pathogens biologically through competition for space by virtue of their ecological obligate association with roots. It is imperative to use native AMF as it has good potential to protect the plant from FOC and not only they protected the host plant but also they influenced their developing nodules and percent recovery of yield loss [55].

11. Conclusion

Fusarium wilt of chickpea undoubtedly remains the grave threat in the consistent production of the chickpea worldwide. This chapter has been considered to discuss the cause of wilt disease and its controlling measures and standard protocol of its extensive evaluation. Since chickpea is a staple crop and to adequately maintain its high yields instantly a synergetic approach is required because in complex nature one preferred method is not sufficient to combat the prevalent disease, therefore, possible attempt has been wisely made to discuss traditional and modern techniques. Biocontrol agents and beneficial AM fungi showed potential to not merely reduce the wilt but also remediate the soil.

Author details

Chandan Singh and Deepak Vyas*
Department of Botany, Lab of Microbial Technology and Plant Pathology,
Dr. Harisingh Gour Central University, Sagar, MP, India

*Address all correspondence to: deepaknvyas64@gmail.com

IntechOpen

© 2021 The Author(s). Licensee IntechOpen. This chapter is distributed under the terms of the Creative Commons Attribution License (http://creativecommons.org/licenses/by/3.0), which permits unrestricted use, distribution, and reproduction in any medium, provided the original work is properly cited.

References

[1] Nene Y, Reddy M, Haware M, Ghanekar A, Amin K, Pande S, et al. Field Diagnosis of Chickpea Diseases and their Control. Int Crop Res Inst Semi-Arid Trop 2012:66.

[2] Patidar R. Studies on Fusarium oxysporum f.sp. ciceri inciting wilt in chickpea (*Cicer arietinum* L.). Rajmata Vijayaraje Scindia Krishi Vishwa Vidyalaya, College of Agriculture,Indore(M.P.), 2017.

[3] Varshney RK, Song C, Saxena RK, Azam S, Yu S, Sharpe AG, et al. Draft genome sequence of chickpea (Cicer arietinum) provides a resource for trait improvement. Nat Biotechnol 2013;31:240-246. https://doi.org/10.1038/nbt.2491.

[4] Singh R, Sharma P, Varshney RK, Sharma SK, Singh NK. Chickpea Improvement: Role of Wild Species and Genetic Markers. Biotechnol Genet Eng Rev 2008;25:267-314. https://doi.org/10.5661/bger-25-267.

[5] Patil MG. Wilt of Chickpea with Special Reference to Characterization of Races and Variant of Fusarium oxysporum f. sp. ciceris and Parameters Associated with Resistance. Jawaharlal Nehru Krishi Vishwa Vidyalaya ,Jabalpur In, 2015.

[6] Nene YL, Sheila YK, Sharma SB. A world list of chickpea and pigeonpea pathogens. 1996.

[7] Singh PK, Singh M, Agnihotri VK, Vyas D. Arbuscular Mycorrhizal Fungi: Biocontrol against Fusarium wilt of Chickpea. Int J Sci Res Publ 2013;3:1-5.

[8] Jiménez-Díaz RM, Castillo P, Jiménez-Gasco M del M, Landa BB, Navas-Cortés JA. Fusarium wilt of chickpeas: Biology, ecology and management. Crop Prot 2015;73:16-27. https://doi.org/10.1016/J.CROPRO.2015.02.023.

[9] Cunnington J, Lindbeck K, Jone R. National Diagnostic Protocol for Fusarium oxysporum f. sp. ciceris The cause of fusarium wilt of chickpea. Subcomm Plant Heal Diagnostics 2016;1:36.

[10] Chand H, Khirbat SK, L CC. Chickpea Wilt and Its Management - a Review. Agric Rev 2009;30:1-12.

[11] Jiménez-Fernández D, Montes-Borrego M, Jiménez-Díaz RM, Navas-Cortés JA, Landa BB. In Planta and Soil Quantification of Fusarium oxysporum f. sp. ciceris and Evaluation of Fusarium Wilt Resistance in Chickpea with a Newly Developed Quantitative Polymerase Chain Reaction Assay. Phytopathology 2011. https://doi.org/10.1094/PHYTO-07-10-0190.

[12] Geiser DM, Ma L-J, Rooney AP, Proctor RH, Manners JM, O'Donnell K, et al. Fusarium Pathogenomics . Annu Rev Microbiol 2013;67:399-416. https://doi.org/10.1146/annurev-micro-092412-155650.

[13] Jendoubi W, Bouhadida M, Boukteb A, Béji M, Kharrat M. Fusarium Wilt Affecting Chickpea Crop. Agriculture 2017. https://doi.org/10.3390/agriculture7030023.

[14] Ahmad N, Lodhi K, Abbas A, Waris W, Asad M, Aslam MM. Chickpea Wilt and its Management Strategies-A Review Paper. Imp J Interdiscip Res (IJIR 2016;2.

[15] Hartman GL, Pawlowski ML, Chang HX, Hill CB. Successful Technologies and Approaches Used to Develop and Manage Resistance Against Crop Diseases and Pests. Elsevier Ltd; 2015. https://doi.org/10.1016/B978-1-78242-335-5.00003-2.

[16] Chaudhry MA, Ilyas MB, Muhammad F, Ghazanfar MU. Sources

of resistance in chickpea germplasm against fusarium wilt. vol. 5. 2007.

[17] Shittu H. FUSARIUM WILTS: AN OVERVIEW. Environ Res J 2012;6.

[18] Bishop CD, Cooper RM. An ultrastructural study of root invasion in three vascular wilt diseases. Physiol Plant Pathol 1983;22:15-IN13. https://doi.org/10.1016/S0048-4059(83)81034-0.

[19] Haware M. . Fusarium wilt and other important diseases of chickpea in the Mediterranean area. Zaragoza: CIHEAM,Options Méditerranéennes : Série A. Séminaires Méditerranéens; n. 9; 1990.

[20] Singh S, Singh I, Kapoor K, Gaur PM, Chaturvedi SK, Singh NP, et al. Chickpea. Broadening Genet. Base Grain Legum., Delhi: 2014, p. 1-215. https://doi.org/10.1007/978-81-322-2023-7.

[21] Ahmad N, Lodhi K, Abbas A, Waris W, Asad M, Aslam MM. Chickpea Wilt and its Management Strategies-A Review Paper. Imp J Interdiscip Res (IJIR) 2016;2:1281-90.

[22] Dubey SC, Singh SR, Singh B. Morphological and pathogenic variability of Indian isolates of Fusarium oxysporum f. sp. ciceris causing chickpea wilt. Arch Phytopathol Plant Prot 2010;43:174-189. https://doi.org/10.1080/03235400802021108.

[23] Arvayo-Ortiz RM, Esqueda M, Acedo-Felix E, Sanchez A, Gutierrez A. Morphological Variability and Races of Fusarium oxysporum f.sp. ciceris Associated with Chickpea (Cicer arietinum) Crops. Am J Agric Biol Sci 2011;6:114-121.

[24] Nath N, Ahmed AU, Aminuzzaman FM. Morphological and physiological variation of Fusarium oxysporum f. sp. ciceri isolates causing wilt disease in chickpea. Int J Environ Agric Biotechnol n.d.;2. https://doi.org/10.22161/ijeab/2.1.25.

[25] Landa BB, Navas-Cortés JA, Jiménez-Díaz RM. Integrated Management of Fusarium Wilt of Chickpea with Sowing Date, Host Resistance, and Biological Control. Phytopathology 2004;94:946-960. https://doi.org/10.1094/PHYTO.2004.94.9.946.

[26] Jiménez-Fernández D, Landa BB, Kang S, Jiménez-Díaz RM, Navas-Cortés JA. Quantitative and Microscopic Assessment of Compatible and Incompatible Interactions between Chickpea Cultivars and Fusarium oxysporum f. sp. ciceris Races. PLoS One 2013;8. https://doi.org/10.1371/journal.pone.0061360.

[27] Yadav D. Studies on wilt (Fusarium oxysporum f.sp. ciceri) of chickpea (Cicer arientinum L.). Rajmata Vijayaraje Scindia Krishi Vishwa Vidyalaya, Gwalior College of Agriculture Gwalior, 2010.

[28] Nene Y., Haware M., Reddy M. Chickpea Diseases: Resistance Screening Techniques, information Bulletins No. 10. Patancheru. Inf Bull No 10 Patancheru, AP, India Int CroDS Res Inst Semi-Arid Trop 1981:1-10.

[29] Silavat S. Studies on Spatial and Temporal Behaviour of Fusarium oxysporum f.sp. ciceri Inciting Wilt in Chickpea Cicer arietinum L. Rajmata Vijayaraje Scindia Krishi Vishwa Vidyalaya, 2015.

[30] Singh V, Santosh K, Lal M, Gupta BK. Fusarium Wilt of Chickpea : An overview. In: Biswas SK, Kumar S, Chand G, editors. Dis. pulse Crop. Their Sustain. Manag., New Delhi: Biotech; 2016.

[31] Upasani ML, Gurjar GS, Kadoo NY, Gupta VS. Dynamics of colonization

and expression of pathogenicity related genes in Fusarium oxysporum f.sp. ciceri during chickpea vascular wilt disease progression. PLoS One 2016;11. https://doi.org/10.1371/journal.pone.0156490.

[32] Gurjar G, Barve M, Giri A, Gupta V. Identification of Indian pathogenic races of Fusarium oxysporum f. sp. ciceris with gene specific, ITS and random markers. Mycologia 2009;101:484-495. https://doi.org/10.3852/08-085.

[33] Sivaramakrishnan S, Kannan S, Singh SD. Genetic variability of Fusarium wilt pathogen isolates of chickpea (Cicer arietinum L.) assessed by molecular markers. Mycopathologia 2002;155:171-178. https://doi.org/10.1023/A:1020479512034.

[34] Sharma KD, Muehlbauer FJ. Fusarium wilt of chickpea: Physiological specialization, genetics of resistance and resistance gene tagging. Euphytica 2007;157:1-14. https://doi.org/10.1007/s10681-007-9401-y.

[35] Chakraborty S, Pangga IB, Roper MM. Climate change and multitrophic interactions in soil: The primacy of plants and functional domains. Glob Chang Biol 2012;18:2111-2125. https://doi.org/10.1111/j.1365-2486.2012.02667.x.

[36] Landa BB, Katan J, Jiménez-Díaz RM. Temperature Response of Chickpea Cultivars to Races of Fusarium oxysporum f. sp. ciceris, Causal Agent of Fusarium Wilt. Plant Dis 2006;90:365-375. https://doi.org/10.1094/PD-90-0365.

[37] Anuga SW, Gordon C. Adoption of climate-smart weather practices among smallholder food crop farmers in the Techiman municipal: Implication for crop yield. Res J Agric Environ Manag 2016;5:279-286.

[38] Singh PK, Vyas D. Biocontrol of plant diseases and sustainable agriculture. Proc. Natl. Acad. Sci. India - Sect. B Biol. Sci., 2009, p. 110-128.

[39] Chaudhary M, Saksena S. Studies on Fungal Metabolites with special reference to the plant growth requlators produced by some fungi. Dr.Harising Gour Vishwavidyalaya, 1973.

[40] Randhawa MK, Saksena S. Studies on Root Nodules Bacteria of Sagar soils. Dr. Harising Gour Vishwavidyalaya, 1970.

[41] Vyas D, Gupta RK. Effect of Edaphic Factors on the Diversity of Vam Fungi. Int J Res Biosci Agric Technol 2014;1:14-25. https://doi.org/10.29369/ijrbat.2014.02.ii.0091.

[42] Dehariya K, Shukla A, Ganaie MA, Vyas D. Individual and interactive role of Trichoderma and Mycorrhizae in controlling wilt disease and growth reduction in Cajanus cajan caused by Fusarium udum. Arch Phytopathol Plant Prot 2015;48:50-61. https://doi.org/10.1080/03235408.2014.882119.

[43] Shukla A, Dehariya K, Vyas D, Jha A. Interactions between arbuscular mycorrhizae and Fusarium oxysporum f. sp. ciceris: effects on fungal development, seedling growth and wilt disease suppression in Cicer arietinum L. Arch Phytopathol Plant Prot 2014;48:240-252. https://doi.org/10.1080/03235408.2014.884831.

[44] Singh PK, Vyas D. To Study Microbial Interaction of Below Ground Organisms With Special Reference to Fusarium wilt ofChickpea. Dr.Harising Gour Vishwavidyalaya, 2007.

[45] Dehariya K, Shukla A, Sheikh IA, Vyas D. Trichoderma and arbuscular mycorrhizal fungi based biocontrol of fusarium udum butler and their growth promotion effects on pigeon pea. J Agric Sci Technol 2015;17:505-517.

[46] Larkin RP, Fravel DR. Efficacy of Various Fungal and Bacterial Biocontrol

Organisms for Control of Fusarium Wilt of Tomato. Plant Dis 2007;82:1022-1028. https://doi.org/10.1094/pdis.1998.82.9.1022.

[47] Dubey SC, Suresh M, Singh B. Evaluation of Trichoderma species against Fusarium oxysporum f. sp. ciceris for integrated management of chickpea wilt. Biol Control 2007;40:118-127. https://doi.org/10.1016/j.biocontrol.2006.06.006.

[48] Prasad P, Kumar J. Management of fusarium wilt of chickpea using brassicas as biofumigants. Legum Res 2017;40:178-182. https://doi.org/10.18805/lr.v0i0.7022.

[49] Sadiq Y, Vyas K. Studies on the Role of Phenolic Compounds in Disease Resistance in Plants. Dr.Harising Gour Vishwavidyalaya, 1988.

[50] Fu ZQ, Dong X. Systemic Acquired Resistance: Turning Local Infection into Global Defense. Annu Rev Of Plant Biol 2013;64:839-863. https://doi.org/10.1146/annurev-arplant-042811-105606.

[51] Iriti M, Varoni EM. Chitosan-induced antiviral activity and innate immunity in plants. Environ Sci Technol 2005;22:2935-2944. https://doi.org/10.1007/s11356-014-3571-7.

[52] Saikia R, Singh T, Kumar R, Srivastava J. Role of salicylic acid in systemic resistance induced by Pseudomonas fluorescens against Fusarium oxysporum f. sp. ciceri in chickpea. Microbiol Res 2003;158:203-213.

[53] Chérif M, Arfaoui A, Rhaiem A. Phenolic compounds and their role in bio-control and resistance of chickpea to fungal pathogenic attacks. Tunis J Plant Prot 2007.

[54] Kumar Y, Zhang L, Panigrahi P, Dholakia BB, Dewangan V, Chavan SG, et al. Fusarium oxysporum mediates systems metabolic reprogramming of chickpea roots as revealed by a combination of proteomics and metabolomics. Plant Biotechnol J 2016;14. https://doi.org/10.1111/pbi.12522.

[55] Shukla A, Dehariya K, Vyas D, Jha A. Interactions between arbuscular mycorrhizae and Fusarium oxysporum f. sp. ciceris: effects on fungal development, seedling growth and wilt disease suppression in Cicer arietinum L. Arch Phytopathol Plant Prot 2015;48:240-252. https://doi.org/10.1080/03235408.2014.884831.